FROM CHILDREN TO ADULTS: HOW BINAURAL CUES AND EAR CANAL IMPEDANCES GROW

Von der Fakulät für Elektrotechnik und Informationstechnik der
Rheinisch-Westfälischen Technischen Hochschule Aachen
zur Erlangung des akademischen Grades einer
DOKTORIN DER INGENIEURWISSENSCHAFTEN
genehmigte Dissertation

vorgelegt von

Diplom-Ingenieurin
Janina Fels
aus Mönchengladbach

Berichter: Universitätsprofessor Dr. rer. nat. Michael Vorländer
Universitätsprofessor Dr.-Ing. Peter Vary

Tag der mündlichen Prüfung: 30. Januar 2008

Diese Dissertation ist auf den Internetseiten der Hochschulbibliothek online verfügbar.

Janina Fels

From Children to Adults:
How Binaural Cues and
Ear Canal Impedances Grow

Logos Verlag Berlin GmbH

Aachener Beiträge zur Technischen Akustik

Herausgeber:
Prof. Dr. rer. nat. Michael Vorländer
Institut für Technische Akustik
RWTH Aachen
52056 Aachen
www.akustik.rwth-aachen.de

Bibliografische Information der Deutschen Nationalbibliothek

Die Deutsche Nationalbibliothek verzeichnet diese Publikation in der
Deutschen Nationalbibliografie; detaillierte bibliografische Daten sind
im Internet über http://dnb.d-nb.de abrufbar.

Dissertation RWTH Aaachen
D 82, 2008

ISBN 978-3-8325-1855-4
ISSN 1866-3052
Band 5

Logos Verlag Berlin GmbH
Comeniushof, Gubener Str. 47,
10243 Berlin
Tel.:   +49 (0)30 / 42 85 10 90
Fax: +49 (0)30 / 42 85 10 92
http://www.logos-verlag.de

Contents

Abstract – Zusammenfassung

Abstract

In this thesis the growth-dependency of head-related transfer functions and their resulting parameters (interaural time and level differences) are analyzed as well as ear canal impedances.

Custom-designed measurements for children were developed which made it possible to obtain anthropometric data from different kinds of subjects, ranging from infants to adults. The obtained data is statistically evaluated with regard to its influence on binaural cues and ear canal impedances with the help of suitable simulation- and measurement techniques.

It becomes evident that children and adults differ tremendously as far as their respective head-related transfer functions are concerned. Furthermore it turns out that the head-related transfer function of a child cannot be obtained by scaling down the dimensions of an adult head. Differences in the anatomy of children and adults thus result in different binaural cues. The individual anthropometric parameters, however, affect the binaural cues to a varying extent. In this thesis the most important anthropometric parameter with regard to their influence on binaural cues, are determined. Ear canal impedances are undergoing certain changes as well while children are growing up and finally reach adulthood. This thesis presents for the first time, the age-related development of data that is most decisive for the impedances.

These new findings open up new possibilities to develop artificial child heads and couplers for hearing aids that are suitable for children. Thus, improvements in certain fields of applications such as the development and fitting of hearing aids for children are now possible and measurement techniques that are used for classroom acoustics can now be optimized significantly as well. Moreover, these new findings are vital when it comes to re-evaluating standardized artificial heads.

Zusammenfassung

In dieser Arbeit werden die Wachstumsabhängigkeiten von Außenohrübertragungsfunktionen und daraus abgeleiteten Größen (Interaurale Zeit- und Pegelunterschiede) sowie von Gehörgangsimpedanzen untersucht.

Für die Untersuchung an Kindern entwickelte Verfahren machen es möglich, die anthropometrischen Parameter vom Kleinkind bis hin zum Erwachsenen zu erfassen. Mittels geeigneter Simulations- und Messverfahren werden diese Daten hinsichtlich ihres Einflusses auf binaurale Merkmale und auf Gehörgangsimpedanzen statistisch ausgewertet.

Es zeigt sich, dass die Außenohrübertragungsfunktion eines Kindes wesentliche Unterschiede zur der eines Erwachsenen aufweist. Ferner kann eine Außenohrübertragungsfunktion eines Kindes nicht durch das Skalieren von Erwachsenen-Abmessungen angenähert werden. Durch die unterschiedliche Detail-Anatomie zwischen Kind und Erwachsenen entstehen große Unterschiede in den binauralen Merkmalen. Allerdings wirken sich die einzelnen anthropometrischen Parameter unterschiedlich stark auf die binauralen Merkmale aus. In dieser Arbeit werden die wichtigsten anthropometrischen Parameter hinsichtlich des Einflusses auf binaurale Größen ermittelt. Auch die Gehörgangsimpedanzen ändern sich stark im Laufe des Wachstums vom Kleinkind bis hin zu Erwachsenen. Die Arbeit zeigt erstmals den Verlauf der altersabhängigen Entwicklung der für die Impedanz maßgeblichen Daten.

Mit Hilfe der gewonnenen Ergebnisse stehen nun neue Möglichkeiten für die Entwicklung von Kinderkunstköpfen und kindgerechten Hörgerätekupplern zur Verfügung. Dadurch können spezielle Kinderanwendungen, wie zum Beispiel die Hörgeräteentwicklung und -anpassung oder auch die Messmethodik in der Klassenraumakustik wesentlich optimiert werden. Ferner dienen die gewonnenen Erkenntnisse der Re-Evaluierung von standardisierten Kunstköpfen.

Introduction

In the last decades, artificial heads (dummy heads, head and torso simulators) have been used increasingly in acoustics for a variety of tasks. They are employed as a directional binaural microphone for measurements that require a clear reference to the sound pressure at the human ear. Examples for applications of the so-called "binaural technology" can be found in audiology, room acoustics, communication engineering, audio engineering and product sound design. As described by FEDTKE ET AL. [FDF⁺07], various applications of binaural technology are well developed. Apart from measurements, sound recording and reproduction are of special interest. The leading pioneers in the field of dummy heads were BURKHARD AND SACHS [BS75] who introduced a head and torso simulator which was intended for hearing aid measurements. From these early days onwards various artificial heads have been developed. In order to make them comparable and to reduce the measurement uncertainties, the anthropometric data of an average human head, torso and pinna are standardized in IEC 60959 [IEC60959] and ITU-T P.57/58 [ITUP.57, ITUP.58], respectively. The same goes for acoustic impedances of the human ear canal and eardrum defined by IEC 60711 [IEC60711] or IEC 60318-5 [IEC60318-5] (former number IEC 126).

When it comes to artificial heads and their applications, one can distinguish two different types.

All linear distortions caused by the torso, head and pinna reflection and diffraction are collected for the first type (cf. Figure 1). In general, the measurement microphone is placed at the entrance of the ear canal, but the ear canal itself is not specifically represented. Such systems are suitable for measuring (distant) sound sources in the far field. All obtained information is hence directional dependent.

The second type is used for sound sources which are close to the ear. Here, the acoustical impedance of the ear canal as well as the replication of the cavum conchae and pinna are important. These parameters are directional independent and can be replicated with so-called ear simulators or couplers attached to the head and torso simulator. The main reason for the need of an ear simulator is that it is necessary to provide the correct radiation impedance for the close-to-the-ear source.

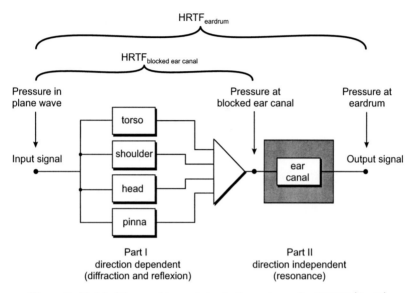

Figure 1: Model of the sound transmission to the eardrum after GENUIT [Gen84]

Both components are taken into account while listening to binaural recordings: the direction-dependent component by the head- and torso geometry and the direction-independent part of the ear canal. A well-defined interface is thus established by separating both components at the ear canal entrance[1].

In recent years, however, criticism has been formulated concerning the applicability of standard dummy heads for an arbitrary group of human population. Standard heads should represent the head and torso of typical adults (see Part I in Fig. 1). One aspect is that the adult anthropometry may have changed since the early days of standardization (see GENUIT AND FIEBIG [GF07] and FASTL [Fas04], for example). Another aspect is related to the statistical variance of individual head dimensions.

In some applications of dummy head measurements such as hearing aid tests, an even more important problem occurs. When focusing on e.g. classroom measurements or hearing aid development for *children* or *infants* the consequences of using anthropometric data of adults apparently remain unknown. First measurements can

---

[1]When listening with headphones to dummy head recordings, one should bear in mind that certain equalizations must be considered. The equalization could be based on headphone transfer functions measured at the individual subjects (cf. [MSHJ95, MHJS99]).

only be done when new measurement heads for children are available.

The anatomy changes while growing up and turning from an infant into an adult. The question is how much and in which systematic way these parameters change. The *Head-Related Transfer Function* (HRTF) describes how a given sound wave input is filtered by the diffraction and reflection properties of the head, pinna, and torso, before the sound reaches the eardrum. During growth, accordingly, this parameter, binaural cues (which can be deduced from the HRTF) and ear canal impedances change as well.

Head-related transfer functions and the ear canal impedance of adults were studied in detail by several researchers (e.g. SHAW [Sha97], MØLLER [MSHJ95], GENUIT [Gen84]). However, no focused study has yet been carried out with regard to the head-related transfer functions of children or infants. This is the starting point of this thesis.

It discusses the influences of the anthropometric data (head, torso, pinna and ear canal) on binaural cues, head-related transfer functions and ear canal impedances. It is studied which parameter causes what kind of binaural cue, which parameter has a big impact and which one has only a small impact. Thus, having obtained the knowledge about the differences between children and adults, the applications of artificial heads as far as children are concerned can be improved significantly. One can even argue that it would be useless to develop hearing aids for children with directional signal processing if these devices are based on measurement that were carried out with adult subjects. Nevertheless the development of hearing aids is only one possible future application for children-sized dummy heads. Classroom acoustics is another obvious field of application.

This thesis describes step by step how head-related transfer functions and ear canal impedances of children and infants can be determined. Moreover, the growth dependency is studied.

An overview of the structure of the thesis is given in Figure 2. Chapter 1 provides the theoretical background and the basics about head-related transfer functions, binaural cues and ear canal impedances. The main contribution of the thesis is then divided into three parts according to the different types of applications of artificial heads.

- Part I deals with the "outer ear (blocked ear canal)". The anatomical parameters up to the point of the ear canal entrance are studied. These parameters have a directional dependent influence, that means that the influence is dependent on the direction of sound incidence due to diffraction and reflection, on the HRTF and the binaural cues. Chapter 2 focuses on the determination of head-related

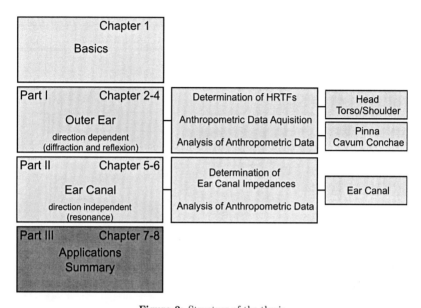

Figure 2: Structure of the thesis

transfer functions. A simulation technique and the ordinary way of measuring head-related transfer functions are compared. How the anthropometric data of the head, torso, and pinna are obtained, is described in Chapter 3. By using a photogrammetric technique it becomes possible to perform a contactless measurement of agile subjects, which is very important for small children, as they keep moving. CAD models, which are used for the simulation, are generated and created pit of this data. In Chapter 4 the influences of the collected anthropometric data on the binaural cues and head-related transfer functions are analyzed. The changes that occur during growth are demonstrated and discussed on a six-months-old child attending in kindergarten and an adult. Then, by varying single parameters separately, the influence of these parameters is identified individually.

- Part II deals with the ear canal. Here, the ear canal parameters and the ear canal impedances are studied. These parameters behave direction independently. In Chapter 5 two methods to determine ear canal impedances, by simulation and by measurement are presented. The growth dependency of the ear canal parameters and its influences on ear canal impedances are discussed in Chapter 6.

- Part III finally summarizes the results with special regard to novel application possibilities for children. The discussion lead us to pilot studies as described in Chapter 7, which show the importance for a new research direction in future. Finally a detailed summary of the thesis is provided in Chapter 8.

Chapter 1

Basics

1.1   The Auditory System

The auditory system is capable of using a great diversity of physical cues. This provides us, amongst others, with the possibility to determine the location of a sound source.

The auditory system encodes the frequency and intensity of sound and extracts information about temporal variation. In addition, the auditory system calculates additional information about differences between sounds arriving at the two ears called interaural differences. Time and level differences between the two ears and differences in the spectral and temporal compositions of sounds due to the reflection and diffraction of head and torso as well as pinna effects have an influence on the subject's perception of the localization of the sound source. As the anatomy changes while growing up and turning from an infant into an adult, the binaural cues and spectral and temporal compositions change accordingly.

In this thesis, we will attach special attention to the outer ear. From an anatomical point of view, the outer ear consists of the head, torso, and pinna and the external ear canal (meatus acusticus externus), which ends at the eardrum (cf. Figure 1.1). The pinna is crucial with regards to spatial hearing. The external ear canal is a slightly curved tube and leads from the cavum conchae to the eardrum. The eardrum is acoustically loaded by the ossicles and the inner ear. The acoustical effect of the pinna, head, and torso is based on reflection, shadowing, dispersion, diffraction, interference, and resonance.

Numerous infant auditory research dealing with the development of hearing in infants (summarized in SAFFRAN ET AL. [SWW06] and LITOVSKY AND ASHMEAD [LA97]) has been carried out, however, most has focused on infants' perception of speech.

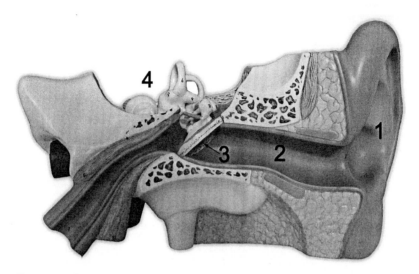

Figure 1.1: Cross section of the ear showing 1: Outer ear (pinna and cavum conchae), 2: External ear canal, 3: Eardrum, 4: Middle- and inner-ear

This work studies the development of hearing in infants from a different point of view. BATTEAU [Bat67, Bat68] was one of the first researchers, who postulated that the external ear, specifically the pinna, could be a source of spatial cues that might account for localization. SHAW AND TERANISHI [ST68] studied the processing of the sound in the pinna using a model of the external ear as well as natural ears. They identified a number of resonance frequencies of the external ear using their model.

1.2 Head-Related Transfer Functions and Binaural Cues

A head-related coordinate system of spherical coordinates according to Figure 1.2(a) is used throughout the thesis to define a frame of reference. This system shifts in conjunction with movements of the subjects' head. The origin of the coordinate system is halfway between the entrances of the two ear canals. The view forward defines the direction with $\varphi = 0°$ and $\vartheta = 0°$, with φ being the azimuth, and ϑ being the elevation. φ rotates counterclockwise and ϑ counts positive at angles in the upper hemisphere. The horizontal plane is thus defined. The frontal plane intersects the entrances of the ear canals and lies at right angles to the horizontal plane. The median plane (also called median sagittal plane) lies at right angles to the horizontal and frontal plane. The three planes intersect at the origin. Thus, the origin lies in the middle of the head.

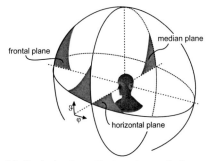

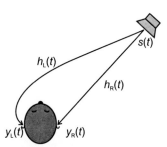

(a) Head-related coordinate system; φ is the azimuth, and ϑ is the elevation

(b) HRIRs for left and right ear describe the filtering of a sound source ($s(t)$) before it is perceived at the left and right ears as $y_L(t)$ and $y_R(t)$, respectively

Figure 1.2: Head-related coordinate system and path of the sound to the ear

As mentioned in the introduction, the *Head-Related Transfer Function* (HRTF) describes how a given sound wave input is filtered by the diffraction and reflection properties of the head, pinna, and torso, before the sound reaches the eardrum.

The linear systems analysis defines the transfer function as the complex ratio between the output signal spectrum and the input signal spectrum as a function of frequency. The transfer function $\underline{H}(f)$ of any *Linear Time-Invariant* (LTI) system at frequency f can be given as:

$$\underline{H}(f) = \frac{\underline{Y}_{\text{Output}}(f)}{\underline{S}_{\text{Input}}(f)}. \tag{1.1}$$

The HRTF $\underline{H}(f)$ is the Fourier transform of the *Head-Related Impulse Response* (HRIR) $h(t)$.

BLAUERT [Bla97] describes three different types of transfer functions:

1. The *free-field transfer function* relates sound pressure at a point of measurement in the ear canal of the experimental subject to the sound pressure that would be measured using the same sound source, at a point corresponding to the center

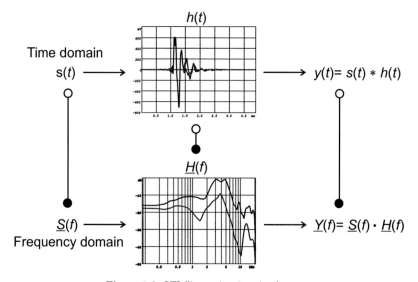

Figure 1.3: LTI (linear time-invariant) system

of the head while the subject is not present.

$$H_{\text{free-field}}\left(f\right) = \frac{H\left(f, r, \vartheta, \varphi\right)}{H_0\left(f, r\right)} \tag{1.2}$$

2. The *monaural transfer function* relates sound pressure at a point of measurement in the ear canal for any given direction and distance of the sound source to sound pressure measured at the same point but with the sound source at a reference angle and distance. (As a rule, a plane wave from the direction $\varphi = 0°, \vartheta = 0°$ is used as a reference.)

$$H_{\text{monaural}}\left(f\right) = \frac{H\left(f, r, \vartheta, \varphi\right)}{H\left(f, r, 0, 0\right)} \tag{1.3}$$

3. The *interaural transfer function* relates sound pressure at corresponding points of measurement in the two ear canals. The reference sound pressure is that at the ear facing the sound source.

$$H_{\text{interaural}}\left(f\right) = \frac{H_{\text{turned away}}\left(f, r_1, \vartheta_1, \varphi_1\right)}{H_{\text{turned towards}}\left(f, r_1, \vartheta_1, \varphi_1\right)} \tag{1.4}$$

When the term HRTF is used in this thesis, it refers to the type 1 HRTF definition mentioned above measured at the blocked ear canal (cf. Fig. 1). An anechoic environment and a great distance to a sparse focusing sound source are the prerequisites for the approximation of a free plane sound field.

The *diffuse-field transfer function* describes the accordant distortions in a diffuse sound field to the eardrum. In this case the incidence of sound waves from all directions have the same probability. Corresponding to an ideal plane wave, the sound pressure level is direction independent in the diffuse field. A diffuse sound field occurs in big rooms beyond the diffuse-field distance[1].

To calculate the *diffuse-field transfer function* it is sufficient to add or average the energies of each solid angle element $d\Omega = \sin\vartheta \, d\vartheta \, d\varphi$. Therefore, a weighting coefficient w_j must be taken into account for each direction dependent HRTF to consider the various sizes of solid angles.

With i = Azimuth-Index and j = Elevation-Index and in case of a measured or simulated database in steps of 5° the diffuse-field transfer function is defined as:

$$G_{\text{diffuse-field}}(f) = 10\log\left[\frac{1}{4\pi}\sum_{j=1}^{37}\left(w_j\sum_{i=1}^{72}\left|\underline{H}_{ij}(f)\right|^2\right)\right] \text{ dB} \qquad (1.5)$$

with the weighting factor

$$w_j = \pi\frac{5°}{180°}\left[\sin\left(\pi\frac{\vartheta_j + 2.5°}{180°}\right) - \sin\left(\pi\frac{\vartheta_j - 2.5°}{180°}\right)\right] \qquad (1.6)$$

and

$$w_1 = w_{37} = \pi\frac{5°}{180°}\left[\sin\left(\frac{\pi}{2}\right) - \sin\left(\pi\frac{92.5°}{180°}\right)\right]. \qquad (1.7)$$

The *directivity index* is calculated according to:

$$d(f) = H_{\text{free-field}}(f) - G_{\text{diffuse-field}}(f) \qquad (1.8)$$

with $H_{\text{free-field}}$ = HRTF in reference direction.

---

[1]That distance from the acoustic center of a sound source at which the mean-square sound pressure of the direct sound, in a specific direction, is equal to the mean-square sound pressure of the reverberant sound in the room containing the source.

Table 1.1: Selection of public HRTF databases

Group	subjects	additional information
MIT Media Lab[1]	1 (KEMAR)	Complete set of HRTF measurements, diffuse-field equalized HRTFs, pinnae in small and "large-red"
CIPIC HRTF Database[2]	45 individuals + KEMAR (small/large pinna)	age, gender, weight + anthropometric data ([ADTA01])
IRCAM[3]	51 individuals	gender, hairstyle
AUDIS project[4]	12 individuals + HMS II resp. HMS III	–
Itakura Laboratory[5]	96 individuals	74 subjects' physical sizes

[1] http://sound.media.mit.edu/KEMAR.html
[2] http://interface.cipic.ucdavis.edu/CIL_html/CIL_HRTF_database.htm
[3] http://www.ircam.fr/equipes/salles/listen/download.html
[4] http://www.eaa-fenestra.org/Products/Documenta/Publications/09-de2
[5] http://www.itakura.nuee.nagoya-u.ac.jp/HRTF/

A multiplicity of databases of individuals or artificial heads has been established. Some public HRTF databases are listed in Table 1.1.

1.2.1 Localization

The spatial hearing of the human auditory system consists out of the detection of the sound direction and distance. Additionally, the human auditory system is able to separate one source out of several spatially distributed sound sources. Numerous (psychoacoustic) experiments have been carried out with regard to the spatial hearing (cf. [Bla97]). It turned out that the azimuth of the source can be determined by using differences in interaural time or interaural level, whichever is present. WIGHTMAN AND KISTLER [WK92] believe that the low-frequency temporal information is dominant if both are present. The elevation of the source is determined with the help of spectral shape cues. The received sound spectrum, as modified by the pinna, head and torso, is in effect compared with a stored set of directional transfer functions. These are actually the spectra of a nearly flat source heard at various elevations. The elevation that corresponds to the best-matching transfer function is selected as the locus of the sound.

1.2.2   Interaural Time and Level Differences

In 1906, LORD RAYLEIGH [Ray07] already provided an explanation for the ability of human sound localization by time differences between the sounds reaching each ear (ITDs) and differences in sound level entering the ears (interaural level differences, ILDs) which is nowadays referred to as the *Duplex theory*. The duplex theory states that ITDs are used to localize low frequency sounds, in particular, whilst ILDs are used in the localization of high frequency sound inputs. Nevertheless the duplex theory has its limitations as the theory does not explain completely directional hearing and as no explanations is given for the ability to distinguish between a sound source directly in front and behind. Furthermore the theory is only suitable when it comes to localizing sounds in the horizontal plane around the head. The theory also does not take into account the use of the pinna in localization. It has also been demonstrated that the Interaural Time and Level Difference is used by humans when it comes to determining source elevation and resolving front-back confusion (MIDDLEBROOKS AND GREEN [MG91]).

The ITD can be approximated by simple formulas (summarized in [MPO+00]), which are deduced from simple geometry models (such as spheres or ellipsoids) of the head. Different methods can be used to calculate the ITD for an individual with the help of HRTFs or HRIRs.

Among others, the detection of the ITD can be accomplished by evaluating the group delay of the excess phase components at 0 Hz, or by detecting the leading edge of the HRIR, or by calculating the maximum of the interaural cross-correlation of the impulse responses. The last method is applied in this thesis. Before cross-correlating the pair of HRIRs (left ear and right ear) the impulse responses can be interpolated several times to obtain a better estimate of the maximum point. Since the ITD is a primary low-frequency localization cue, the impulse responses can be low-pass filtered before the cross-correlation takes place. The ITD calculated in this thesis are low-pass filtered at 1500 Hz edge frequency before the cross-correlation.

The calculation of the ILD is based on level differences of the HRTF spectra (of the left and right ear) which are averaged across frequency. However, the ILD is strongly frequency-dependent. In this thesis the broadband ILD is calculated for reasons of simplicity.

Several studies determined the *just noticeable difference* (JND) for the ITD and ILD for adults. The JND of the ITD is approx. $10 \, \mu s$ ([Bla97] p. 153) and for the ILD approx. $1.5 \, dB$ ([Bla97] p. 161).

1.2.3 Artificial Heads

As a consequence of the growing interest in headworn hearing instruments, Knowles Electronics developed the *KEMAR Manikin* (Knowles-Electronics Manikin for Audiologic Research) as a tool for improving the measurement and reporting of the performance of hearing aids in the early seventies [BS75, Bur78]. Since its first description and introduction in 1972, all hearing aid manufacturers, numerous research audiologists and others have studied the characteristics of hearing aids mounted on the artificial head.

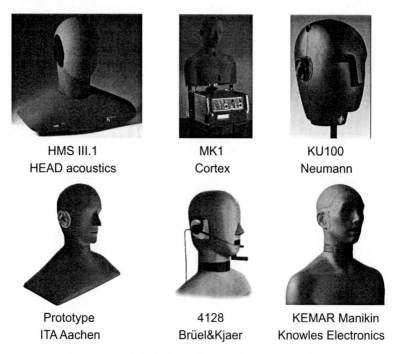

Figure 1.4: Artificial heads (Source: FEDTKE ET AL. [Fed07])

Application of Artificial Heads

The KEMAR was designed with median human adult dimensions. Ear simulation matches the acoustic response with an auricle, an ear canal, an eardrum that equals the median ear in dimension, acoustic impedance and modes. Dimensions of torso and head are based on anthropometric data, but the auricle was based on data obtained especially for this purpose. The ear canal and eardrum are adapted from the ear like coupler by ZWISLOCKI [Zwi70, Zwi71] (cf. Section 1.3.2). Validating measurements have proved the KEMAR to be like a median human in acoustic response to free-fields. In particular head and body diffraction effects are encountered, what became very important when hearing aids with directional microphones were introduced.

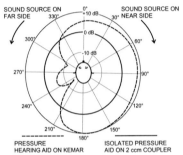

(a) Directivity pattern of an omni-directional hearing aid on the right ear of KEMAR in comparison to free-field

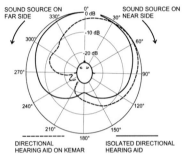

(b) Directivity pattern of a gradient microphone hearing aid on the right ear of KEMAR in comparison to free-field measurements

Figure 1.5: Directivity pattern of hearing aids on the right ear of KEMAR in comparison to free-field ($f = 2\,\text{kHz}$) (by courtesy of Siemens Hearing Solutions)

Driven by the development of head worn directional hearing aids, the hearing aid designer needed to place the microphone inlets in order to optimize directionality. Since the sound field on different positions on or near the ear is important for *Behind The Ear* (BTE) hearing instruments, the introduction of the KEMAR was a milestone for the hearing health care. Due to diffraction effects on the head, a natural directivity, as known from the head-related transfer functions, is observed. Nevertheless, during typical BTE measurements the microphone is placed on top of the ear instead of in the ear, while the receiver of the hearing instrument is acoustically coupled to the ear coupler (see Section 1.3.2) representing the rest volume of the closed ear channel. Assuming a directional BTE hearing instrument with microphone positions aligned just above the right ear, the directivities of the head and the directional microphone add. Figure 1.5 compares free-field and KEMAR measurements of the directivity patterns of a directional hearing aid with a gradient microphone system. Considering the nose position of the head being the reference direction, the directivity pattern of the head worn hearing instrument is de-arranged to the side due to the head diffraction. A shift of the directivity pattern to the front direction would be desirable to achieve as much spatial gain in the signal to noise ratio as possible for the hearing impaired. This can be achieved by an optimized position of the microphones. A proper placing of the microphones is crucial even for modern digital hearing systems. Figure 1.6(a) shows, for instance studies of hearing instruments featuring a second order directional microphone. In this case, the sensitivity to amplitude and phase distortions produced by the head is much higher than for gradient microphones. In Figure 1.6(b) the directivity pattern obtained with a second order directional microphone built in a hearing aid is displayed at a frequency of 2 kHz. The comparison to a very optimized first order directional microphone points out the benefit for the hard of hearing people keeping out noise from backward and side directions.

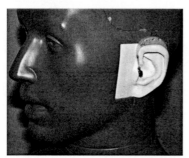 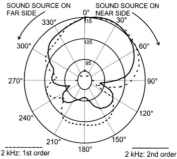

(a) Digital hearing instrument with second order directional microphone on KEMAR

(b) Directivity pattern of hearing instruments with first and second order directional microphone on KEMAR

Figure 1.6: Digital hearing instrument with second order directional microphone on KEMAR (by courtesy of Siemens Hearing Solutions)

Traditional measurement techniques are still very common as far as hearing aids are concerned (IEC 118, [IEC60118-0, IEC60118-7, IEC60118-8]). Due to the strong non-linear nature of modern signal processing techniques used in hearing instruments, measurements are performed under quasi-static conditions using sinusoidal excitation signals. Although correlation- or speech-related measurement approaches are sometimes used to characterize different aspects of the hearing aid more objectively, the traditional way of hearing aid testing is generally still used nowadays. Controlled by reference measurements which are carried out separately in free-field conditions without head, the sound pressure level of the exciting sound field is kept constant at the head center position. The manikin is placed in this equalized sound field to measure the total acoustic performance of the head worn hearing instrument. Thus, artificial head measurements are a substantial part of the evaluation and development of modern hearing systems.

However, artificial heads are applied in other fields of interest dealing with children, too. Applications in the field of room acoustics (especially when children are concerned – classroom acoustics) are only one possibility for further applications besides many other.

1.3 Ear Canal Impedances

The ear canal (meatus acusticus externus) is a tube running from the outer ear to the middle ear (cf. Fig. 1.1). The ear canal extends from the pinna to the eardrum. A typical adult ear canal is about 2–3 cm long and has a diameter of 0.6–0.8 cm. The ear canal is slightly convex curved in the back and upper direction (for protection). The outer half to two thirds of the canal is surrounded by cartilage and contains glands that produce *cerumen* (ear wax), while the inner third to half is surrounded by bone.

The general definition of the field impedance $\underline{Z}_s$ is the ratio between the sound pressure $\underline{p}$ and the particle velocity $\underline{v}$

$$\underline{Z}_s = \frac{\underline{p}}{\underline{v}} \tag{1.9}$$

$[Z_s] = \mathrm{Ns/m^3} = \mathrm{kg/(m^2 s)}$.

The acoustic impedance $\underline{Z}_a$ is defined as the ratio of the sound pressure $\underline{p}$ and the volume velocity $\underline{q}$

$$\underline{Z}_a = \frac{\underline{p}}{\underline{q}} \tag{1.10}$$

with

$$\underline{q} = \int \underline{\vec{v}}\, \mathrm{d}\vec{A} \tag{1.11}$$

$[\underline{Z}_a] = \mathrm{Ns/m^5} = \mathrm{kg/(m^4 s)}$; $[\underline{q}] = \mathrm{m^3/s}$. In this thesis the ear canal impedance is defined according to Eq. (1.9).

1.3.1 Importance of the Ear Canal Impedance

There are many applications for the impedance of the ear in the field of medicine. The ear canal impedance is very important when it comes to fitting hearing aids and checking the hearing ability.

Detection of Hearing Impairment of Children and Adults

Hearing impairment can either be detected subjectively or objectively. The hearing ability of adults and older children, is tested by means of audiometric tests. Audiometric tests determine a subject's hearing level with the help of an audiometer. Several kinds of audiometric tests exist, for example pure tone audiometry, speech audiometry or Békésy-audiometry. However, children or infants are not able to give any feedback about their hearing, especially very young infants are not able to do so. Thus, various objective methods estimate the hearing ability at very early stages.

The impedance audiometry determines how the middle ear works. It does not tell whether the child is hearing or not, but it helps to detect any changes in pressure in the middle ear. A small probe with a microphone and loudspeaker is inserted into the ear and at a certain tone (usually 220 Hz) the sound pressure is measured in front of the eardrum. Most part of the sound is directed by the eardrum into the middle ear, while a small part is reflected. The ratio between the reflected part and the inserted sound depends on the resistance of the eardrum and the middle ear behind. If an alteration of the eardrum flexibility occurs, this is registered by this measurement method. On the basis of this measurement, the function of the middle ear can be tested. This kind of measurement provides only a rough estimation, since only a certain frequency ([LNCJE93]) is tested and just the imaginary part is evaluated ([SWM02]). It is difficult to carry out this test when young children are involved, as the child needs to sit very still and should not cry, talk or move at all.

Many different types of hearing tests can be used to evaluate the child's hearing. Some of them may be used for all age groups, while others are used for a specific age group and a certain level of understanding. There are two primary types of hearing screening methods for newborns. *Evoked otoacoustic emissions* (EOAE) (see Figure 1.7) - a test that uses a tiny, flexible plug that is inserted into the baby's ear. Sounds are sent through the plug. A microphone in the plug records the otoacoustic emissions (responses) of the normal ear in reaction to the sounds. There are no emissions if the baby suffers from a hearing loss. This test is painless and is usually completed within a few minutes, while the baby is sleeping.

Auditory brainstem response (ABR) audiometry (or *Brainstem electric response audiometry* (BERA)) is a neurologic test of the auditory brainstem function in response to auditory (click) stimuli. ABR audiometry refers to an evoked potential generated by a brief click or tone pip transmitted from an acoustic transducer in the form of an inserted earphone or headphone, while the baby is sleeping. The elicited waveform response is measured by surface electrodes typically placed at the vertex of the scalp and ear lobes. Thus, the test measures the brain's activity in response to the

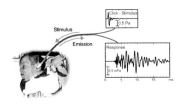

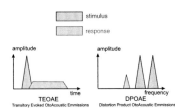

(a) Measurement procedure of otoacoustic emissions

(b) Measurement procedure of distortion product otoacoustic emissions

Figure 1.7: Measurement procedure of (distortion product) otoacoustic emissions

sounds. As in EOAE, this test is painless and takes only a few minutes. If the screening tests detect that the child suffers from a hearing loss, further testing is needed. It is recommended that all babies with hearing loss are identified by three months of age, so that treatment can begin before the baby is six months old as this is an important time for speech and language development.

Applications of Ear Canal Impedances – Fitting of Hearing Aids

The ear canal impedance plays an important role as far as the fitting of hearing aids is concerned, if the impedance is not only evaluated at one single frequency [LNCJE93]. Additionally KEEFE [KBAB93] postulates that there is a direct connection between the values of the growth dependent differences in the ear canal impedances and studies of the organ of hearing.

Usually the hearing aid is placed with an ear mold in the entrance of the ear canal and then a small tube is sending the sound into the ear canal. The goal is to provide a correct sound pressure at the eardrum.

In general three methods can be used when fitting hearing aids. The first one is the "in situ" measurement which means that a small probe microphone is placed in front of the eardrum in addition to the hearing aid in order to test the individual transmission characteristics of the hearing aid and the otoplastic.

The second method is a measurement featuring a test box and a coupler (see next section). The hearing aid is attached to a coupler which should reproduce the ear canal volume and impedance. Thus, the hearing aid settings can be tested without the patient. The standard IEC 60118-7 [IEC60118-7] stipulates the use of the 2 cm^3 coupler according to IEC 60318-5 (Revision of IEC 60126:1973) [IEC60318-5] for measurements with test boxes.

As a combination of this two methods, the *RECD-Measurement* (Real-Ear to

Coupler Difference) developed by MOODIE ET AL. [MSS94] measures the difference
(in dB) as a function of frequency between dB(SPL), measured in a 2-cm³-coupler
(see next section) and in an individual´s ear canal, using the same source of sound
pressure.

1.3.2   Coupler

Couplers (or sometimes called "ear simulators" or "artificial ears") are used as a (standardized) replica of the ear canal and impedance. One field of application is the
fitting of hearing aid. The ear canal volume and impedance is reproduced using several cavities and the pre-settings of hearing aids can be calculated. Such standardized
couplers (as well as standardized artificial heads) can only provide a "mean ear", so
that individual properties cannot be considered. However, for children a standardized
coupler is not available.

The BELL COUPLER is one of the first couplers and was developed by the company
Bell at the end of the 1930s. It was supposed to be used for tests of the speech quality
in telephones. This coupler is a copy of a realistic ear canal and made out of natural
rubber (see Figure 1.8(a)). On the one hand this coupler was thus very good, but on
the other hand the natural rubber made it impossible to produce repeatable results.
The shape of the rubber ear is not stable and ageing was an additional problem.

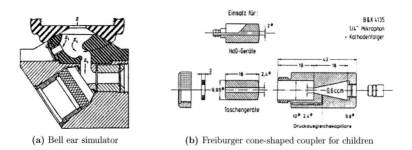

(a) Bell ear simulator   (b) Freiburger cone-shaped coupler for children

Figure 1.8: Two different couplers

ZWISLOCKI proposed in 1970/71 [Zwi70, Zwi71] a coupler for earphone calibration. This coupler consists of one main cavity and four additional cavities, which are
adjustable. The ZWISLOCKI COUPLER has been accepted as national standard S3.25
[ANSIS3.25] by the American National Standards Association.

In the field of hearing aids this 2 CM^3 COUPLER is the most famous and widespread coupler. This coupler is standardized in the IEC 60318-5 (Revision of IEC 60126, 1973) [IEC60318-5]. This coupler is used when measurements are carried out in accordance with IEC 60118 [IEC60118-0]. The construction of this coupler is very simple. It essentially consists of a cylindrical cavity whose compliance is to that of a volume of 2 cm^3. The coupler is made of hard, dimensionally stable, non-porous materials. However, this coupler does not reproduce the human ear very well since the cavity of this coupler is too large compared to a real human ear. This is especially true for children. It turned out that the deviation of RECD-corrected data from standard 2 cm^3-coupler-data is about 15–19 dB(SPL) for infants (cf. [Ric80]). For reasons of robustness and production tolerances, however, this coupler is still in use.

The ear simulator according to IEC 60711 [IEC60711] is an improvement with regard to the reproduction of the ear canal. This ear simulator, that has similar characteristics to the ZWISLOCKI COUPLER, was introduced in Europe by Brüel&Kjær. This coupler has a volume of 1.57 cm^3, which represents the average volume of an adult, which is 1.26 cm^3 better. The distinctive feature of this coupler is, that certain acoustic properties are defined in the reference plane. Therefore, the acoustic specifications at the reference plane are described. Figure 1.10 shows an ear simulator (Wideband Ear Simulator Type 4195) created by Brüel&Kjær. Figure 1.9(b) shows the electrical equivalent diagram. This coupler is not used in practice due to the very complex construction and the cost involved.

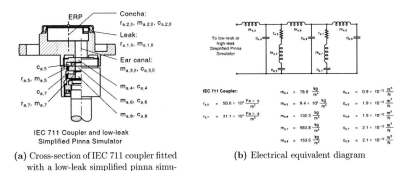

(a) Cross-section of IEC 711 coupler fitted with a low-leak simplified pinna simulator

(b) Electrical equivalent diagram

Figure 1.9: Cross-section of IEC 711 and electrical equivalent diagram (Source: Brüel&Kjær)

The only existing coupler for children is the FREIBURGER KONISCHE KINDER

KUPPLER (FKKK - a cone-shaped coupler for children made in Freiburg by KELLER [Kel85a, Kel85b]). The volume of this coupler is 0.6 cm^3 and it is intended for children who are between three and five years old. The geometrical data of this coupler is based on a study by PFEIL [Pfe79]. For the first time, it is possible, when using this coupler, to fit hearing aids for children without the help of correction factors. Figure 1.8(b) shows the cross-section of the coupler.

Part I

Outer Ear (Blocked Ear Canal)

Chapter 2

Determination of Head-Related Transfer Functions

Basically, head-related transfer functions can be determined using two different techniques. In this chapter, first this two methods will be described and their the pros and cons are weighed up. The first, and more common method is the direct approach namely the measurement of the head-related transfer functions on individuals. The second alternative is the indirect approach to obtain the Head-Related Transfer Function (HRTF). This indirect approach can be split in two ways. However, for the indirect approach exact anthropometric data which describe the head and torso are needed. How this data can be obtained is described in detail in Chapter 3. As soon as the anthropometric data is collected, a *Computer Aided Design* (CAD) model of the head and the torso can be created. This model can either be used to fabricate for instance a wooden dummy head, or to apply numerical methods to calculate the HRTF.

2.1 Measurement

The measurement of HRTF is the most common way to obtain individual HRTFs. In this case the test subject is usually sitting on a chair in an anechoic environment. Figure 2.1 shows in (a) the measurement of an artificial head. In this case the head is placed on a turn table and a pivot arm is moving the loudspeaker to the desired positions. The subject sits down on a chair and measurement microphones (probe-microphones or subminiature microphones) are placed in the entrances of the (blocked) ear canals to measure individual human HRTFs (see Figure 2.1(c)). Figure 2.1(b) shows an individual measurement of HRTFs in the horizontal plane. The question whether to use blocked meatus or not is addressed by HAMMERSHØI

AND MØLLER [HM96]. They concluded that blocked-entrance head-related transfer functions have a lower standard deviation between subjects than open entrance and eardrum HRTFs. Additionally, the introduction of the blocked entrance makes it possible to divide the total transmission from the free field to the eardrum into a directional-dependent part (Part I in this thesis), and a directional-independent part (Part II in this thesis), in a way that renders the two parts uncorrelated.

The measurement signal is usually a broadband deterministic signal such as a sweep. The position of the subjects' head needs to be fixed during the measurement. Measuring the HRTF in the horizontal plane in steps of 5° takes about 20 minutes. Depending on the future application, the HRTF measurements are desired in the complete upper hemisphere with a very detailed directional resolution.

  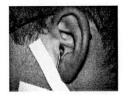

(a) Measurement of the head-related transfer function of an artificial head

(b) Measurement of the head-related transfer function of an individual

(c) Probe Microphone in the ear canal

Figure 2.1: Measurement of head-related transfer functions on an artificial head and on individuals

A reference measurement is required to measure the free-field HRTF (according to Eq. (1.2), p. 10) as well. The (subminiature) microphones are placed into the origin (center of the head, with the head being absent) to carry out the required reference measurement. Thus, the influence of the distance between the measurement loudspeaker and the microphone is excluded as well as the influence of the measurement loudspeaker and microphone themselves. The main advantage of the individual measurement is that very individual results can be obtained which are of great importance when it comes to localization tests. Additionally, the measurement method provides a high resolution in the frequency range (typically covering the range of 20 Hz to 20 kHz).

In contrast to adults, children are not able to take part in such a measurement. Apart from the awkward situation in the anechoic chamber, the main challenge is to remain motionless during the measurement. Thus, the individual measurement of HRTFs is not used for this thesis. However, this measurement technique must be applied, too, if the first of the indirect approaches is used. The microphones are placed in the entrances of the ear canals of the dummy head, which is created according to the anthropometric data of an individual or the mean values provided by a database (cf. measurement of an artificial head, Figure 2.1(a)). This method is applied later in Section 2.3 to verify the simulation results.

2.2 Simulation

The second indirect approach is the numerical simulation of HRTFs. The *Boundary Element Method* (BEM) is a well-known numerical procedure to calculate the radiation of a closed, finite surface. The surface, in this thesis the head and torso model, needs to be described by its geometry and boundary conditions. For this thesis, the full 3D models are generated using MicroStation (BENTLEY[1]) in combination with the software I-deas[2].

Some considerations are necessary to carry out the numerical simulation efficiently and to minimize the calculating time. In correspondence to the natural hearing sensation and to the ordinary measurement procedure as described in Section 2.1, the sound source is assumed to be in the far field, and the eardrum acts as the receiver. The direct simulation of this method would require an immense computer capacity as every angle of sound incidence has to be simulated. This situation, however, can be avoided by a reciprocal arrangement (cf. [Str73], [Lya59], [Sha76], [Fah95]). The pressure caused at the ear canal p_1 by a source of volume velocity Q_2 at position 2 is equivalent to the pressure p_2 at point 2 caused by a source of volume velocity Q_1 at the ear canal. This is due to the reciprocity of the GREEN'S FUNCTIONS (Eq. (2.3)). Figure 2.2 and Equation (2.1) illustrate this relation.

$$\frac{p_1}{Q_2}\bigg|_{Q_1=0} = \frac{p_2}{Q_1}\bigg|_{Q_2=0} \qquad (2.1)$$

Thus, a circular vibrating area in the ear canal is used as a piston source instead of the loudspeaker. The normalization to incident plane waves is described in the section below. For this thesis the blocked meatus (ear canal entrance) is chosen as the reference point because it offers many advantages, especially as a clear reference condition. The results can be compared and discussed more easily since influences

---

[1]MicroStation: A suite of CAD software products for 2- and 3-dimensional design and drafting. http://www.bentley.com/en-US/Products/MicroStation/Overview.htm

[2]I-deas: Integrated Design and Engineering Analysis Software. http://www.plm.automation.siemens.com/en_us/products/nx/

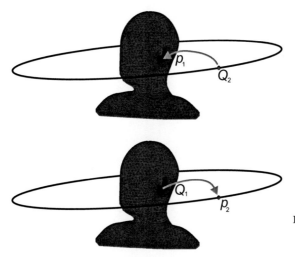

Figure 2.2: Reciprocity
in the determination
of the HRTF

from the ear canal can be excluded. Another advantage is the well-defined interface
for an artificial ear (cf. Part II).

Q, in the reciprocal arrangement, is calculated out of $v_{\text{source}} \cdot S_{\text{source}}$, with S_{source}
equals the surface area of the blocked ear canal entrance ($S_{\text{source}} = 4\pi \cdot r^2_{\text{ear canal entrance}}$)
which is very small, compared to wave-length we discuss and v_{source} equals the normal
velocity. So the Q is like a point source on a surface (cf. KATZ [Kat01] and KAHANA
ET AL. [KNPC98]). The validity of this approach is proofed in the next section.

The boundary surface of a CAD model needs to be discretized into a triangular
mesh to apply the BEM (see Figure 2.3(a)). An important aspect is that the BEM
data matrix needs to be completely stored in the cache memory of the CPU to avoid
extensive computational operations of data storage management. A further limitation
is the fact that the computation time increases with the sixth power of the frequency.
A certain number of computation points on the surface are needed to obtain accurate
results. Six points per wavelength are needed at least in the regions of high sound
pressure for linear interpolations [Zie77]. The maximum distance between two nodes
is thus defined as $d_{max} = \lambda/6$. In this thesis the mesh-sizes of the models vary
depending on the size of the model and on how detailed they are. The number of
nodes vary from approx. 2100 to 6800. The torso and the head are meshed with
elements of length approx. 1 cm. In areas with a more detailed geometry (i.e. the
ear) a very fine mesh is chosen (less than 0.5 mm). Accordingly, the upper frequency

in the results obtained is chosen with 8 kHz for this thesis. This frequency range covers the area of normal speech which ranges from 100 Hz to 7 kHz. This range is used since creating hearing aids is one of the most important applications of an artificial head for children.

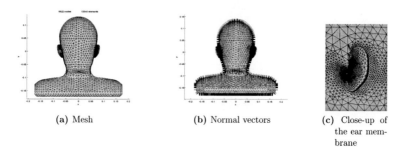

(a) Mesh   (b) Normal vectors   (c) Close-up of
                                                        the ear mem-
                                                        brane

Figure 2.3: Meshing

Having applied the mesh elements on the surface, the normal vectors of the elements can be checked Fig. 2.3(b). The boundary conditions must be defined for all nodes or elements of the mesh. The surface admittance, the ratio of pressure and normal velocity, and a distribution of vibration on the surface acting as excitation must be defined. In this thesis the surface admittance of the head and torso model is defined as 0, which corresponds to a rigid surface.

BÉKÉSY [Bék32] postulated that the acoustic impedance of skin over bone equals the impedance of the surface of water. The characteristic impedance of water is $1.48 \cdot 10^6$ kg/m²s. That means that the characteristic impedance of the skin is 3570 times higher than the characteristic impedance of air. It can be concluded that the skin is rigid. This is confirmed by KATZ [Kat00] who determined an average reflection coefficient of $\overline{|R|} = 0.97$ for skin and postulated that skin behaves rigid.

A hard surface is used to compare the results obtained while using heads made of wood or any other common material in use for producing artificial heads. The impedance of skin could be taken into account in further studies on individual heads as well. At the moment though, this is considered to be of secondary importance. Corresponding to the reciprocal arrangement, the ear canal entrance area vibrates with the normal velocity of 1 m/s. A close-up of the vibrating area of the ear canal entrance and the surrounding area is plotted in Figure 2.3(c).

The problem described here is known as one of the standard problems of numerical acoustics: the radiation problem ([Mec02]). Although it is always better to come to an analytical conclusion with a general character, this is, however, not possible for arbitrary geometries defining the radiation problem. A couple of formulations to solve exterior field problems numerically have been developed. In the following, the most popular formulation based on the KIRCHHOFF-HELMHOLTZ INTEGRAL EQUATION is presented representatively for all similar formulations. This formulation, called direct form or sometimes basic form, is also used in this thesis. The KIRCHHOFF-HELMHOLTZ INTEGRAL EQUATION is

$$\iint\limits_{S} \left[p(y)\frac{\partial g(x,y)}{\partial n(y)} - \frac{\partial p(y)}{\partial n(y)}g(x,y) \right] \mathrm{d}s = \begin{cases} p(x) & ,x \in B_e \\ C(x)p(x) & ,x \in S \\ 0 & ,x \in B_i \end{cases} \qquad (2.2)$$

with

$$g(x,y) = \frac{1}{4\pi\tilde{r}}\mathrm{e}^{(-\mathrm{j}k\tilde{r})} \qquad (2.3)$$

$$\tilde{r} = \|x - y\| \qquad (2.4)$$

$$C(x) = 1 + \frac{1}{4\pi}\iint\limits_{S} \frac{\partial}{\partial n(y)}\left(\frac{1}{\tilde{r}(x,y)}\right) \mathrm{d}s(y). \qquad (2.5)$$

S describes a closed surface, B_e is the exterior field room, B_i is the interior field room, y is a spatial point on the surface, x is the location of a field point. $C(x)$ is the solid angle seen from x. Eq. (2.2) simply says that the pressure at any field point can be calculated by integrating monopoles and dipoles located on the surface. Unfortunately this distribution is not known a priori. The pressure distribution on the left side of (2.2) has to be calculated from these boundary conditions. This is the task of the BEM. The second line of Eq. (2.2) also gives the correct relation of surface pressure to monopole and dipole distribution on the surface. As the pressure on the surface is on both sides of the equation, one has to solve this equation for the surface pressure.

This formula has to be discretized and the integration has to be done numerically to be able to use it for computations. This causes some problems since the numerical integration is not possible for singular integral kernels as it is the case here. Besides, the equation suffers from the so-called non-uniqueness problem. Both problems are described and solved in literature and therefore not worth to be discussed in this context. A detailed description of the BEM and these specific problems can be found, for instance, in [Mec02]. The non-uniqueness problem is discussed comprehensively in BURTON AND MILLER [BM71]. Methods of discretization and interpolation on surfaces are described in [Zie77].

The discretized form of the KIRCHHOFF-HELMHOLTZ INTEGRAL EQUATION for the surface pressure is

$$\sum_{k=1}^{N} \iint\limits_{S_k} \left[p(y)\frac{\partial g(x_i, y)}{\partial n(y)} + \mathrm{j}\omega\rho_0 v_n(y)g(x_i, y) \right] \mathrm{d}s = C(x_i)p(x_i), \qquad (2.6)$$

with the normal velocity

$$\frac{\partial p(y)}{\partial n(y)} = -\mathrm{j}\omega\rho_0 v_n(y). \qquad (2.7)$$

The integration is split into a sum of integrations over the N surface elements. In the example shown in Figure 2.3 the elements are planar-triangular.

A strategy to avoid numerical problems is the so-called CHIEF POINT METHOD (combined integral equation formulation method) (Chapter by OCHMANN [Mec02]). A research code of the Institute of Technical Acoustics (RWTH Aachen University) programmed in MATLAB is used for the BEM processing. The pressure is solved for frequencies from 100 to 8000 Hz in steps of 100 Hz in this thesis. One simulation takes three to four hours on a standard PC with 3.2 GHz CPU and 2 GiB RAM depending on the model.

The distribution of the potential of the boundary is the result of the BEM. The magnitude (zoomed into the ear canal entrance, where the vibrating area is) and the phase of the potential at 3 kHz is shown in Figure 2.4.

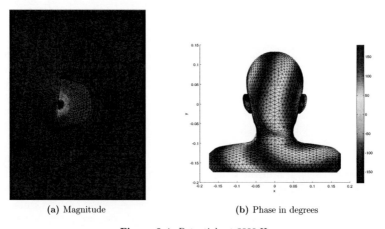

(a) Magnitude   (b) Phase in degrees

Figure 2.4: Potentials at 3000 Hz

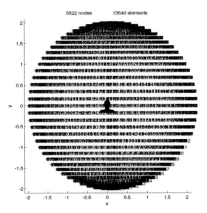

Figure 2.5: Fieldpoints in the postprocessing

Field points can be created and the sound pressure in these field points can be calculated during further steps of post-processing. This means that the sound pressure can be calculated rather quickly for all points of a receiver surface.

Figure 2.5 shows a typical field point distribution to calculate a complete set of HRTFs in steps of 5° in horizontal and azimuthal direction in a distance of 2 m. This field point distribution is used for the following simulations in this thesis.

Calculation of the Head-Related Transfer Functions (HRTF)

Consequently, the head-related transfer functions can be computed according to Equation (2.8) based on the BEM simulation results.

$$\text{HRTF}\,(\vartheta,\varphi) = \frac{p\,(r,\vartheta,\varphi)}{j\omega\rho_0 Q \mathrm{e}^{-\mathrm{j}kr}/4\pi r} \qquad (2.8)$$

With HRTF (ϑ,φ) denoting the head-related transfer function for a specific direction in polar angular direction ϑ and azimuthal direction φ, $p\,(r,\vartheta,\varphi)$ the sound pressure obtained at the far field point in the specific direction and distance r, ω the angular frequency, ρ_0 the equilibrium density of air, Q the volume velocity of the radiation area at the ear canal entrance and k the wave number. The denominator represents the difference between sound radiation and sound reception in terms of the low-frequency law, thus correcting for the radiation impedance in the reciprocal approach (as for instance in [BV04]).

A sphere with a diameter of 16 cm (corresponding to a human adult head) was created to prove the theorem that the approach of a small vibrating area as a source in the reciprocal arrangement is valid (with Q is calculated out of $v_{\text{source}} \cdot S_{\text{source}}$). This sphere was created to prove the accuracy using various vibrating areas. The origin of the coordinate system is identical with the origin of the sphere. Three spheres were built with a vibrating surface area on the 90° side (corresponding to the ear canal entrance) for the test arrangement. The first sphere was built with a vibrating surface area of 8 mm in diameter. The second sphere has a vibrating area of 4 mm in diameter and the third has 2 mm in diameter.

The membrane is disretized into very fine mesh elements. A certain velocity is assigned to each node. The excitation velocity is set to 1 m/s in the simulation However, the excitation is defined on the surface areas of the elements and then converted to the nodes. Each node belongs with regard to the surface areas to the neighboring elements. This approach yields a numerical excitation which is equivalent to the actual surface area of the membrane.

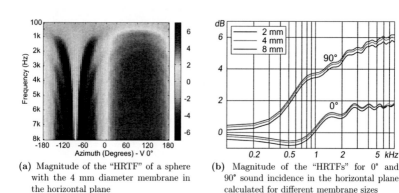

(a) Magnitude of the "HRTF" of a sphere with the 4 mm diameter membrane in the horizontal plane

(b) Magnitude of the "HRTFs" for 0° and 90° sound incidence in the horizontal plane calculated for different membrane sizes

Figure 2.6: Magnitude of a sphere with various membranes

If Q is now calculated out of $v_{\text{source}} \cdot S_{\text{source}}$, an offset is detected, however, very small. If the numerical excitation was equivalent to the actual surface area of the membrane (which occurs when all nodes are assigned to the same velocity), the actual surface area of the membrane would become larger. This causes an offset in the result, since the actual Q is larger. However, if only relative results were regarded this would be negligible, but in terms of absolute values there is an error left. A complete set of "HRTFs" is calculated according to Equation (2.8) in a distance of 2 m for the three spheres.

Figure 2.6 shows the magnitude of the "HRTFs" for the frontal incidence and incidence from 90°.

The three different membranes cause a very small shift in the level of the magnitude of the "HRTF" (less than 0.2 dB). Very small differences of less than 0.2 dB for all angles of incidence for frequencies up to 8 kHz can be put down to the membrane size as well. These differences also depend on the quality of the mesh and the size of the mesh elements. While comparing these results with measurement uncertainties in the ordinary measurement with loudspeaker and microphones and other effects like skin impedance or hair [Rie03], these differences can be neglected. The membranes for the various calculated heads described in this thesis are of 4 mm in diameter.

2.3   Comparison of Simulation and Measurement

The very same head can be simulated and measured to verify the simulation with an ordinary measurement method. In addition to the simulation two models are measured to demonstrate the approach. Therefore the CAD models were created using the measured dimensions from adult test subjects by GENUIT [Gen84] (see Figure 2.7). On the one hand the CAD model is meshed and simulated with BEM and on the other hand the CAD model is used as a basis to build a wooden head with a molding cutter. Wood is used because it is relative easy to handle and almost rigid. The wooden head is measured in an ordinary way. Microphones are placed in exactly the same position as the vibrating area in the simulation. Thus, the measurement procedure is carried out in the ordinary way as described in Section 2.1 and the simulation is carried out in the reciprocal arrangement. The measurement resolution is equivalent to the simulation results, namely 5° in azimuth and elevation. When it comes to the head after GENUIT the finest mesh elements of the head and torso (except pinna) were chosen to 1 cm. Figure 2.7 shows the comparison of simulation and measurement results. In the simulation the vibrating area was applied in the left ear canal entrance. The results of the measurement of the left ear are shown in this figure. Figure 2.7(a) shows the magnitude for sound incidence of 0° (upper figure) and 60° (lower figure) and accordingly Fig. 2.7(b) shows the phase responses. A very satisfying concordance as far as the magnitude as well as the phase response is concerned, is achieved. All |HRTF|-values are plotted in one Figure to check the quality of the simulation as far as the most interesting plane, the horizontal plane is concerned. Figure 2.7(c) shows the |HRTF|-values for the simulated and measured head. The colorbar indicates the |HRTF|-values in decibels (dB). Additionally, the difference between these values are plotted in a zoomed scale of ±10 dB. If the simulation would yield exactly the same results as the measurement this plot would be completely green (equals to 0 dB). Some small differences up to 2 dB can be observed in frequencies up to 6 kHz. However, above 6 kHz larger differences can be observed which might be caused by the quality of the mesh.

The second comparison between simulation and measurement is carried out with a head with the dimensions of a typically six-year-old child (how these dimensions are obtained is described in detail in Section 3.1 (p. 42) and more details about this specific head can be found in Section 4.4 (p. 75)). The CAD-model was discretized with elements smaller than 1 cm, especially in areas around the ear (element length here approx. 0.3 mm). This yields very satisfying results, which are shown in Figure 2.8. The magnitude of the HRTF is almost identically ((a), (c)). Only very slight differences in the phase response (b) at frequencies above 4 kHz can be detected.

All in all, one can say that the method of BEM yields very satisfying results that can be processed fast for frequencies up to 8 kHz. If results at higher frequencies needs to be solved more RAM and computation time will be needed.

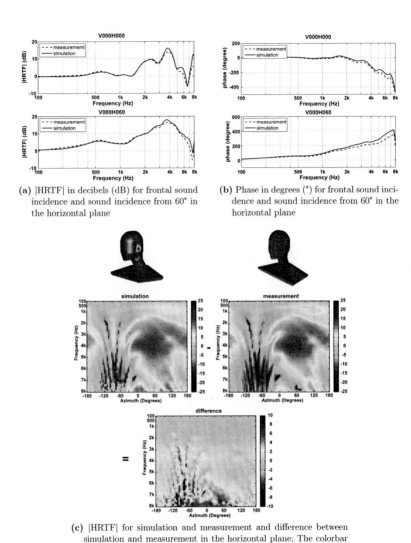

(a) |HRTF| in decibels (dB) for frontal sound incidence and sound incidence from 60° in the horizontal plane

(b) Phase in degrees (°) for frontal sound incidence and sound incidence from 60° in the horizontal plane

(c) |HRTF| for simulation and measurement and difference between simulation and measurement in the horizontal plane; The colorbar indicates the |HRTF|-values in decibels (dB)

Figure 2.7: Comparison between simulation and measurement; Simulation is carried out with a maximum distance between two nodes to achieve good results up to 6 kHz ($d_{max} = \lambda/6$)

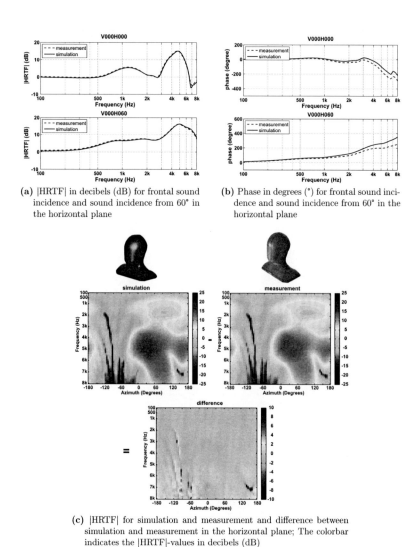

(a) |HRTF| in decibels (dB) for frontal sound incidence and sound incidence from 60° in the horizontal plane

(b) Phase in degrees (°) for frontal sound incidence and sound incidence from 60° in the horizontal plane

(c) |HRTF| for simulation and measurement and difference between simulation and measurement in the horizontal plane; The colorbar indicates the |HRTF|-values in decibels (dB)

Figure 2.8: Comparison between simulation and measurement; Simulation is carried out with a maximum distance between two nodes to achieve good results up to 8 kHz $(d_{max} = \lambda/6)$

Chapter 3

Anthropometric Parameters and CAD Modeling

The main goal of this thesis is to analyze the effects that anthropometric parameters describing the head, torso and pinna, have on binaural cues. Chapter 2 describes a user-friendly and efficient method to obtain these cues with the help of CAD models. Models of this kind can either be created as a copy of an individual human being or with the help of a certain number of values, which are based on statistical evaluation of anthropometric parameters. There are a several possibilities to create both *individual* CAD models and/or *parametric* ones. The best known and most efficient methods are listed in Table 3.1.

Table 3.1: Selection of the best known and most efficient methods to create head and torso models

method	type of model
anthropometer	parametric
laser scanner	individual, parametric
plaster cast	individual
photogrammetry	individual, parametric

So-called **anthropometers** are used for the oldest method. Figure 3.1 (a) and (b) show typical anthropometers which are used to measure the different parameters of a human being. This method is used in many historic studies. It is very cost-effective, but also very time consuming. Besides, this method is strenuous for very young children, since a lot of time is needed to obtain a complete dataset, which is detailed enough to create CAD models. The modeling of an individual human being is impossible. A statistical analysis, however, can be carried out and with the help of this database models can thus be created.

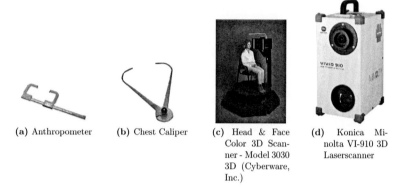

| (a) Anthropometer | (b) Chest Caliper | (c) Head & Face Color 3D Scanner - Model 3030 3D (Cyberware, Inc.) | (d) Konica Minolta VI-910 3D Laserscanner |

Figure 3.1: Various Systems and methods to obtain anthropometric data

Using **laser scanner** is probably the best known method in this field of interest. The advantages of this method are obvious. The laser scanner produces results fast and very accurately. An exact copy of the individual human being is formed. This method provides good results for adults and in addition statistical data evaluation is possible. The disadvantages of these methods are: they are very expensive; a very fast scanner is required for children since the children need to remain still during measurement; parents are afraid that the laser might damage the children's eyes. Additionally, the construction of a complete CAD model without any free edges turned out to be another serious challenge. It is easy to create parts of the head, but fitting the parts together correctly for one closed model, which is needed for the computation, is fairly problematic. Another problem is that the laser scan produces a very fine scan with a high resolution. If the model is used for the BEM calculation, this scan needs to be altered, because a very high fine structure would cause an immense computational load. Moreover, a laser scanning system is not mobile unless it is a very expensive one. Figure 3.1 (c) and (d) shows two laser scanner which are commonly used to obtain anthropometric measurements. This method allows the CAD modeling of an individual as well as the creation of a database to provide values for a parametric model.

Making a **plaster cast** of the head and the pinna is a method which may be reasonable when it comes to one or two individual models, but the amount of effort required, exceeds the practical applicability.

Another possibility to obtain the anthropometric data is the **photogrammetry**. Stereo-photographs, which are used to measure the anthropometric data, are needed for this method (details of this method see Section 3.1). This method has the following advantages: it is accurate (depending on the experience) and it is very safe for the subject, since only photographs are taken from the subjects. This method is best suitable for children, as they do not have to keep still for a long time. Additionally to the modeling of an individual being, statistical data evaluation is possible as well. Furthermore, photogrammetry is cost-effective. The disadvantage is, that this method is very time consuming in the post processing (no automated process, needlework). The accuracy depends on the experience of the researcher and time.

Previous surveys by other researchers have been carried out with different aims in mind. Table 3.2 shows a selection of the most comprehensive surveys. RANDALL ET AL. (1946) [RDBP46] were among the first to use body measurements to derive sizing schemes for flight equipment after many fitting problems occurred during World War II. Anthropometric sizing became routine business for military organizations in several countries. The other relevant measures were estimated from means and standard deviations for the men who fell within an interval. Similarly, ZEIGEN ET AL. (1960) [ZAC60] used head circumference intervals to divide the anthropometry into sizes for helmets.

The datasets, where the Knowles Electronics Manikin for Acoustical Research (KEMAR) (see Section 1.2.3) is based on [BS75, Bur78], are carried out by DREYFUSS (1967) [Dre67] ALEXANDER (1968) [AL68] and CHURCHILL (1956) [CT56]. More than 4000 American subjects were measured for the three studies

One further study, who leads to another artificial head was carried out by GENUIT (1984) [Gen84]. He measured six male adults and made a statistical evaluation out of the approx. 40 values who describe head and pinna.

Two databases containing the anthropometric data of children are really famous. PRADER ET AL. 1989 [PLMI89] carried out a long-term study about the physical growth of Swiss children from birth to 20 years of age from 1954 to 1976. From 1972 to 1977 SNYDER published [SSOVE72, SSOS75, SSO+77] a very comprehensive study about the anthropometric data of children from aged 2 weeks up to 13 years. This database, called *Anthrokids* contains individual and statistical data based on more than 4000 children. Figure 3.2 shows a selection of the head dimensions obtained in this study. Unfortunately these databases were not created with regard to the development of artificial heads, as the values do not refer to the ear canal entrance. Hence, a reconstruction of a head with pinna and correct ear canal entrance point is not possible.

Table 3.2: Selection of databases of anthropometric data containing head and torso measurements

survey	subjects	age/gender	additional information
Dreyfuss, Alexander, Churchill 1956-1967	> 4000 (American) subjects	adults male/female	initial data for KEMAR
Genuit 1984	6 subjects (German)	male adults	statistic data with focus on development of artificial heads
Prader 1989	274	birth–20 years male/female	statistic data, longitudinal study (1954–1976)
Synder[1] 1972-1975	4027 children	2 weeks–13 years	statistic and individual data
CAESAR[TM][2]	2.400 north American, 2.000 European	18–65, male/female	statistic data
Japanese Body Dimensions Data[3] 1997-98	200	18–29, male/female	individual data
International Standard IEC 60959, ITU p.58	norm data	adults	norm, statistic data
DIN 1978, 2005	6000 (German) adults	18–65	norm, statistic data
WEAR Group[4]			World Engineering Anthropometry Resource

[1] http://ovrt.nist.gov/projects/anthrokids/
[2] http://store.sae.org/caesar/, http://www.hec.afrl.af.mil/HECP/Card4.shtml
[3] http://www.dh.aist.go.jp/research/centered/anthropometry/index.php.en
[4] http://ovrt.nist.gov/projects/wear/

The databases described above were all created using conventional methods such as anthropometers. Another two, more recent studies are certainly worth mentioning. On the one hand there is the CAESAR[TM] database, containing measures of 2400 North American and 2000 European aged from 18–65 and on the other hand the

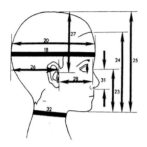
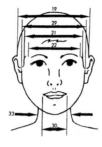

Figure 3.2:
Anthropometric measurements of the study of SNYDER [SSOVE72, SSOS75, SSO+77]

Japanese Body Dimensions Data containing individual data of 200 18–29 years old adults. These databases were created by using laser scanners as well.

An international group, called WEAR (World Engineering Anthropometry Resource), was founded to create a world wide resource of anthropometric data for a wide variety of engineering applications.

For the sake of completeness the normative data is listed here. The International Standard IEC 60959 TR:1990 *Provisional head and torso simulator for acoustic measurements on air conduction hearing aids* [IEC60959] and the ITU-T P.58 RECOMMENDATION *Head and Torso Simulator for Telephonometry* and ITU-T P.57 RECOMMENDATION *Artificial Ears* [ITUP.58, ITUP.57] describe the exact data and tolerances for standardized artificial heads for adults.

How the anthropometric data of the German population has changed during the period from 1978 to 2005 can be observed in the DIN 33402-2 *Ergonomics - Human body dimensions* where the old and new data is compared [DIN33402-1, DIN33402-2].

All in all, there are various methods to obtain the anthropometric measures and there have been comprehensive surveys collecting these measures. However, with regard to children and infants, none of the existing data can be used (as mentioned above, the values do not refer to the ear canal entrance). The advantages and disadvantages as far as this method is concerned, need to be considered carefully. As this thesis focuses on the growth dependency, a method has to be used, that is appropriate for the problematic issue of dealing with children and infants. Photogrammetry is thus the method of choice, because it is the most comfortable procedure for children, as only photographs need to be taken. In addition, the parents' consent is required for such a study and the safer the method, the greater the participation that can be expected (which is another reason not to use a laser scanner).

3.1 Photogrammetry – A Method to obtain Anthropometric Data

For this thesis photogrammetric measurements are carried out to obtain the anthropometric data, which is needed to create CAD models of various heads. Photogrammetry is a measurement technology where the three-dimensional coordinates of points on an object are determined by measuring two or more photographic images taken from slightly different positions.

3.1.1 Basics of Photogrammetry

Photogrammetric techniques have been used for more than 150 years to extract spatial information from 2D pictures. This technique allows a reconstruction of objects and the identification of object features from photographs without any direct contact with the objects. Results of photogrammetric analyzes may be

- coordinates of objects in a 3D coordinate system (digital identification of points)

- drawing (analogue or digital), namely maps or ground plans

- photographs, above all rectified photographs, which provide the basis for aerial maps

Photogrammetry is a measurement procedure, which was originally used in geodesy. The main fields of application are cartography and land surveying. The photographs are rectified and serve as the basis for maps. These are the tasks of the aerial photogrammetry and remote sensing. Additional fields of application such as architecture, engineering, police investigation, measurements of deformation emerged with the construction of photogrammetric cameras for the near field (distances of 1 m – 100 m to the object) and especially with the development of stereoscopic cameras.

The equipment for the photogrammetric data acquisition is portable, as the data sets for this thesis were obtained in kindergartens, day-care centers and primary schools. Figure 3.3(a) shows the portable setup in a gymnasium of a kindergarten. Two standard commercial digital cameras are mounted on a horizontal bar in a distance of approx. 50 cm to each other on a tripod. The cameras are turned with a slight angle to each other to focus on one object in a distance of 1.20 m. The children sit on a chair with an attached height-adjustable headrest (see Fig. 3.3(b)). This chair stands on a turntable, so that the child can be turned from a frontal view to a view from the back. Therefore the turntable is equipped with two locking positions. If the children are very young, the father/mother can sit on the chair holding the infant on his/her lap. The angle is chosen carefully so that the frontal photographs and photographs from the back intersect in a certain area. Thus one half of the head can be photographed in stereo vision at different perspectives without changing light and camera position.

The cameras are manually, synchronously released. This is necessary, since stereo photographs of lively children are made. If the children remain rather calm, the manually release may cause a very small time shift which can be neglected. A minimal time shift, however, prevents the use of flashes. Besides, very young children might be afraid of a flash. Therefore an indirect source of lightening with reflector panels made of polystyrene is used. This light is much more agreeable for the subjects.

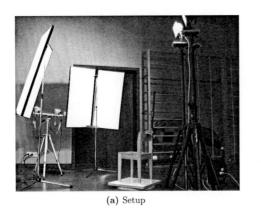

(a) Setup   (b) Turntable

Figure 3.3: Set-up for the stereophotos

Common points need to be identified on each image. Therefore the control points are located in the images and additional tie points are determined (for a detailed description of this procedure see Section 3.1.3). The control points are small black circles with a diameter of 3 mm. There is another white circle with a small black dot in the middle of these circles. Figure 3.5 (right) shows some of these control points. These points are printed on an adhesive film. Since these points need to be fixed on the children, this film is attached on an adhesive hypoallergenic non-woven fabric. This plaster is put onto the skin and onto a swimming cap. The cap is used, as hair has a very fine structure which causes problems when it comes to evaluating in stereo vision. Furthermore the swimming cap presses the hair tightly on the head, so that the shape of the head is much more visible.

Overall four photographs are taken: two digital photographs from the front and two from the back. Figure 3.4 shows for an example the stereo photographs of a six-months-old subject. The two images on the left show the stereo shot from the front and two images on the right show the stereo shots from the back. In this study only one half of the head was photographed. Some parts of the backside of the upper pinna cannot be seen in stereo view; however, for this basic study, these parts are not relevant. The side of the head, which was used, was chosen at random. It is well known that faces are not completely symmetric. The deviations from the left-right symmetry and corresponding cues for localization are certainly important in studies of binaural hearing of individuals. However, to make things easier, these differences are neglected, which led subsequently to a symmetrical database for artificial heads similar to the data in [IEC60959]. A systematic difference can be excluded and since the selection of the head side is a randomized process, a statistical evaluation on this basis is admissible.

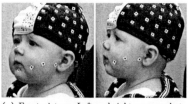 

(a) Front picture: Left and right camera shot   (b) Picture from the back: Left and right camera shot

Figure 3.4: Stereophotographs from a six-months-old child

3.1.2 Calibration of the Cameras

Before the children can be photographed, the digital cameras need to be calibrated. First each camera is tested of its distortion and the values are stored in a special file. A calibration test object is photographed to obtain the lens distortion. The Equations (3.1) and (3.2) calculate the lens distortion. Thus the deviation between the effective image rays and the model of central projection is determined.

$$d_x = A_1 \left(r^2 - r_0^2 \right) x + A_2 \left(r^4 - r_0^4 \right) x + A_3 \left(r^6 - r_0^6 \right) x + B_1 \left(r^2 + 2x^2 \right) + B_2 2xy \quad (3.1)$$

$$d_y = A_1 \left(r^2 - r_0^2 \right) y + A_2 \left(r^4 - r_0^4 \right) y + A_3 \left(r^6 - r_0^6 \right) y + B_1 2xy + B_2 \left(r^2 + 2y^2 \right) \quad (3.2)$$

with d_x and d_y denote the correction factors of the lens distortion, r equals the image radius (distance from the image point to the image center $r^2 = x^2 + y^2$), r_0 equals the zero crossing of the distortion, A_{1-3} denote the polynomial coefficients of the radial symmetric distortion and $B_{1,2}$ equal the coefficients of the tangential distortion.

Second, since an uncalibrated stereo-camera is used here, the angle of the cameras to each other and the exact position is calibrated. For that reason a special test object is photographed. This object consists of two plates mounted in an angle of 90°. Figure 3.5 shows this test object. The object has many control points (small black circles with a diameter of 3 mm). This figure shows a close-up picture of the control points, too. This object is photographed in different positions and angles by rotating the object on the turntable.

Thus each camera takes a picture of the object from its specific viewpoint. A special software is used to calibrate the cameras and the images of the children. PHIDIAS - The Photogrammetric Interactive Digital System - is a digital photogrammetric system which is used for any kind of image measurement (see [BE91, BS97]). PHIDIAS is integrated into MicroStation (by BENTLEY) and suited for 3-dimensional facilities.

The implemented automatic target measurement is used to detect the signs with a very high accuracy and reliability. The center of the sign is the center of the contour-ellipse or -circle which separates the dark shape from the bright background. If the sign has a diameter of approx. 10 pixels, the accuracy which can be achieved is about 1/5 to 1/20 of the pixel size (standard deviation of the image coordinates). The user only has to move the cursor near the sign or can start an automatic interactive or non-interactive measurement of all known points if approximate values for the orientation parameters and object coordinates exist. As far as the test object is concerned, the automatic non-interactive measurement of all known points is carried out. All of the signs are measured for each view and with the help of this data the exact position of the cameras to each other, can be calculated. Figure 3.6 shows the calculated positions of the cameras for five stereo-photographs (red) and the photographed calibration object (black).

3.1.3 Image Orientation and Stereo Measurement

Before taking the photographs the object points have to be marked with the target marks and a sufficient number of images with a good geometrical constellation have to be supplied. The single points need to be measured in each photograph, and each point is assigned to a number. In case of the children each marker has to be

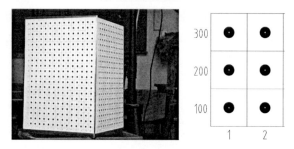

Figure 3.5: Test object with calibration markers

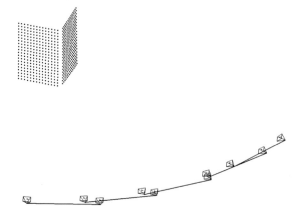

Figure 3.6: Calibration: Calculation of the exact positions of the cameras

measured manually (see Figure 3.7 (a)). After measuring the single points, the three-dimensional coordinates of the object points are calculated by means of the PHIDIAS-bundle adjustment. Prior to the evaluation the image orientation for each stereo set of the children (Fig. 3.4) must be conducted. The orientation program of PHIDIAS performs a complete bundle adjustment of all observations and is provided with all necessary modules to obtain approximate values. Every single step of the orientation (relative orientation, absolute orientation) is carried out under the interactive control of the user. Partial results such as accuracy or the dimension of gross errors are indicated.

The mathematical model of the photogrammetry is based on the collinear equations (Eq. (3.3) and (3.4)).

In a complete bundle adjustment the following unknown parameters are calculated:

- Interior orientation for each camera

  - focal length (principal distance)
  - principal point coordinates x_h, y_h
  - radial symmetric distortion A_{1-3}
  - tangential distortion $B_{1,2}$

- Exterior orientation for each image

  - projection center coordinates (X_0, Y_0, Z_0)
  - angle of rotation (omega, phi, kappa)

The collinear equations (Eq. (3.3) and (3.4)) describe the central projection of 3-dimensional positions into a 2-dimensional image plane. The measured image coordinates depend on the unknown parameters. The sums of squares of the deviations between the measured and calculated image coordinates are minimized in the bundle adjustment.

$$x' = x_0' - c\frac{r_{11}(X - X_0) + r_{21}(Y - Y_0) + r_{31}(Z - Z_0)}{r_{13}(X - X_0) + r_{23}(Y - Y_0) + r_{33}(Z - Z_0)} + dx' \qquad (3.3)$$

$$y' = y_0' - c\frac{r_{12}(X - X_0) + r_{22}(Y - Y_0) + r_{32}(Z - Z_0)}{r_{13}(X - X_0) + r_{23}(Y - Y_0) + r_{33}(Z - Z_0)} + dy' \qquad (3.4)$$

with X, Y, Z equal the 3-dimensional coordinates of a point, x_0' and y_0' denoting the principal point on the CCD, c equals the focal length and r_{ij} is a 3×3 matrix calculated out of the angles of rotation (omega, phi, kappa).

In combination with LCD shutter glasses the stereo images can be evaluated three-dimensional. Two views are used for the left and right stereo image which will be transformed automatically so that no parallaxes - even with convergent directions - will be seen. Figure 3.7 (b) shows this procedure with a six-months-old child. The three-dimensional cursor is controlled by mouse and keyboard so that even elements with curved surfaces can be evaluated. All MicroStation tools and commands can be used for the evaluation. The superimposed drawing will of course be seen in stereo as well.

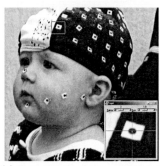

(a) Determination of the marker in the stereo photographs

(b) Three-dimensional evaluation with the help of shutter glasses

Figure 3.7: Determination of the marker in the stereo photographs

3.1.4 Definition of a Head-Related Coordinate System in the Stereo Photographs

First of all a head-related coordinate system must be defined in order to create CAD models on the basis of the stereo photographs or to make precise measurements which refer to the coordinate system which is introduced in Chapter 1 (Figure 1.2) in the photographs. The coordinate system has its origin in the middle of the head on the axis between the ear canal entrances. Only the skin is visible in the stereo photographs and this requires an indirect procedure to determine the center of the head.

Figure 3.8 shows all steps of the procedure. First of all the ear canal entrance point is defined. (On the photograph this point seems to lie on the skin of the ear, but the ear canal entrance can only be seen on the photographs from the back. Hence, the dot seems to be on the skin.) Figure 3.8(a) shows additionally how the x-direction of the coordinate system is defined. A line is placed through the center of the eyes (pupils). The midsagittal plane is defined through the tip of the nose and a marker which is placed in the middle of the chin. The eye-line is then copied to the ear canal entrance point and extended until a rectangular cross section is reached with the midsagittal plane (see Fig. 3.8(b)). This point of intersection is the origin of the head-related coordinate system. In (c) an auxiliary coordinate system is introduced which needs to be turned so that the y-axis points to the top of the head and the z-axis points to the $(0, 0, 0)$ point.

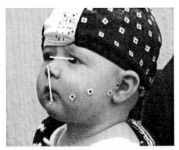

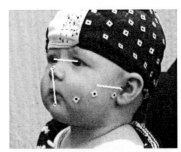

(a) Definition of the x-direction, midsagittal plane and ear canal entrance point

(b) Extension from ear canal entrance in x-direction until midsagittal plane

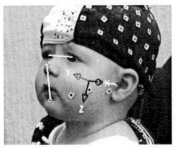

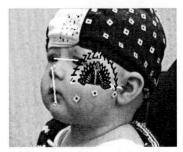

(c) Definition of an orthogonal coordinate system

(d) Rotation of the coordinate system

Figure 3.8: Definition of the head-related coordinate system in the stereo photographs

3.2   CAD Modeling

CAD models of the head and torso can be created in two ways with the help of photogrammetric tools and evaluation. One possibility is to build a model which represents the exact copy of an individual head. The result would be similar to a laser scanned model. The construction of a parametric abstract CAD model based on measured dimension from the head is the second alternative. A good parametric abstract model on the basis of the dimensions of one individual should yield ideally very similar results compared to the individual model.

3.2.1   Individual Model vs. Parametric Model

Figure 3.9 shows the technique of the individual modeling. In the stereo view smart lines are drawn on the skin of the face and swimming cap. Then the smart lines are transformed into B-splines (Fig. 3.9(a)). Hence the surface is generated out of the B-splines (Fig. 3.9(b, c)). In case of a face with no fine structure (nose, eyes, lips etc.), the distances between the splines can be generously chosen depending on the required accuracy. When it comes to the pinna the splines must be very close together to rebuild the fine structure of the ear.

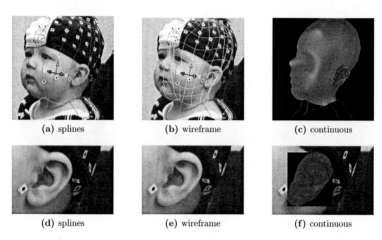

(a) splines   (b) wireframe   (c) continuous

(d) splines   (e) wireframe   (f) continuous

Figure 3.9: Individual model

Creating a model like this, yields certain problems that can be compared to those of the laser scanner method. Parts of the head are easy to create. But the merging, for example, to create one closed model is rather complicated, since the fine structure of the ear yields very complex and highly curved surfaces, which need to be stitched up with simple curved surfaces from the face. Additionally, the meshing of a very detailed pinna is also difficult as the distance between two nodes cannot be infinitesimal.

That is why this procedure is not suitable for fast calculation and modeling. However, to calculate the HRTF for only a few individuals or to prove the accuracy of an abstract parametric model this method can be employed.

Since the detailed realistic model yields almost the same results as an ordinary measurement (verified in Section 2.3), this detailed model is used in the following to prove the accuracy of an abstract parametric model for children.

3.2.2   Determination of a Parametric Model

First of all the parameters have to be selected to create a parametric model of the head. A good model should represent the original adequately, however, the number of parameters is limited. Every measure (length, angles, radii) can be obtained with the help of the photogrammetric method. Using these values, a model can be created.

Very simple models, used for the research of binaural hearing, are based on a sphere with the diameter of a human head or on a so-called snow man model. Spherical and ellipsoidal models of the torso combined with a classical spherical-head model are used for these models ([AAD01a, ADD⁺02]). GENUIT [Gen84] created an abstract parametric model with the help of six datasets consisting of the anthropometric data of six male adults. Figure 3.10 shows the anthropometric data GENUIT measured and the resulting CAD model which was created on the basis of mean values of six male adults (b).

Creating a CAD model is fairly simple, since this model is reduced to only a few parameters. Furthermore, the ear is simplified using planar surfaces and circles.

The simplified (geometric) pinna is sufficient for frequencies below 6-8 kHz (cf. [Sha66]). Above these frequencies the fine-structure of the pinna becomes relevant.

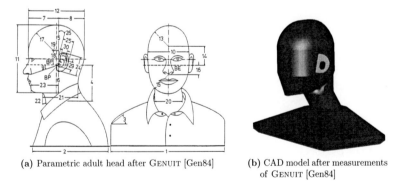

(a) Parametric adult head after GENUIT [Gen84]   (b) CAD model after measurements
                                                             of GENUIT [Gen84]

Figure 3.10: Anthropometric Parameters for an adult (cf. GENUIT [Gen84])

However, this model is used in the first step to create an abstract model of the six-months-old child of the previous figures. Nevertheless, measuring the corresponding dimensions of a six-months-old child, led to a model head which is too rough and only an approximation. The limited number and the improper choice of the free

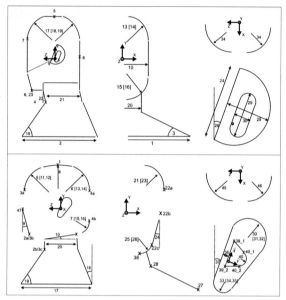

Figure 3.11:
Anthropometric
Parameters for adults
(top) and children
(bottom)

parameters of this model are the reason for this failure (described in detail in FELS
ET AL. [FBV04]). The simplified model features many simplifications which are
appropriate for an adult head. The radii for example (in [Gen84] No. 17 and No. 34,
see Figure 3.10 and 3.11) on the front and the back upper head are assumed to be
equal, which is a good approximation for adults. The head of a six-months-old child,
however, has extremely different radii.

The CAD model was modified to build an abstract parametric model which is
similar to a child.

Many changed were made in the model. Figure 3.11 (lower part) shows the new
model which is appropriate for children. For example the adult model in the yz-plane
the head is approximated with straight lines. But a child's head is rounder than an
adult's head. Important changes must be made in the neck and in the back of the
child's head. Little children do not have a distinctively long neck like adults. The
shape of the pinna and cavum conchae cannot be adequately approximated by an
ellipse as well.

Figure 3.12 (a) shows the optimized parametric head. If the detailed and the
optimized parametric models are overlaid, only very small differences in the geometry

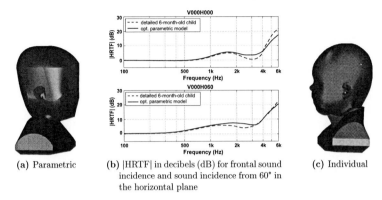

(a) Parametric   (b) |HRTF| in decibels (dB) for frontal sound   (c) Individual
incidence and sound incidence from 60° in
the horizontal plane

Figure 3.12: Individual six-months-old child and optimized parametric child head and
simulation results

remain. Figure 3.12(b) shows the magnitude of the frontal HRTF of the optimized
simplified head and the detailed head. A very good concordance was achieved. Thus,
this improved model makes it possible to create simplified models of heads and torsos
representing the geometries of test subjects (children and adults) to achieve a proper
correlation with the actual anatomy. The adult parameter set presented by GENUIT
[Gen84], nevertheless, is a sub-set of the children parameter set.

3.3   Database

The anthropometric parameters according to Figure 3.11 for the child are measured
in the stereo photographs using the principles described in Section 3.1. A special
script calculates automatically the radii, the vertexes in the specific directions and
the distances required and stores these values in a database. Radii of the head are
determined as follows: a spline is drawn on the skin as close as possible to the required
plane in the coordinate system. Then, the spline is leveled into the plane. Then a
circle is created through the point on the spline to approximate the spline as good as
possible, since splines cannot be evaluated statically. Afterwards the script calculates
the radii, the center of the circles and the vertexes.

For the upper radius of the head, for example, the following data are calculated:

1. radius

2. center: x, y and z coordinate

3. radius: vertex $\pm$ y, $\pm$ z

The photographs have been obtained in kindergartens, schools, and courses for children and adults. First of all a kindergarten is chosen where this procedure will be tested in terms of its practical application. In Germany, children in kindergartens are between three and six years of age. At the beginning the parents were informed in detail about the study, as they had to give their consent. Although, this method does not hold any risks for the children and the parents were promised additionally that the photographs will not appear anywhere and that the data will be recorded anonymously, and that the photographs would be taken during the normal school time, so that the parents would not have to spend any extra time for this procedure, only less than 50 % of the parents gave their consent.

The children behaved very straightforward. Their birthdays and day of the photographs were registered, as well as the sex, the body height, the head circumference and the width of the shoulders. Afterwards the children put on the swimming cap with the markers on it and took a seat on the chair of the turn table.

It turned out that this procedure works excellent even with young children. However, children under the age of three need their parents for this procedure. Thus, the largest age group in this thesis is the group of children attending kindergartens. Older children were photographed in elementary schools and high schools. In toddler groups younger children aged from 6–12 months have been photographed, too.

Thus, in total 95 subjects are registered and manually dealt with. Figure 3.13 shows the relation between the body height and the age for all subjects. The blue triangles represent male subjects and the red squares represent female subjects. A strong connection between age and body height appears up to the age of six years. Furthermore, no statistical difference between male and female subjects can be observed up to the age of eight years. Between eight and 15 years girls seem be a little taller than the boys of the same age. Above 15 years it is the very opposite. A similar trend was observed in [BR88].

Figure 3.14 shows the head circumference of all photographed subjects. This measure is obtained with an ordinary tape measure and not with the photogrammetric system. Fig. 3.14(a) shows the circumference versus the age – Fig. 3.14(b) shows the circumference versus the body height. The circumference increases fast in the first four years fast and grows linearly above four years of age. If plotted versus body height, the circumference increases more linearly. In this group a slight difference between male and female subjects can be observed.

The following Figure 3.15 depicts several head and torso parameters as a function of age. These functions roughly describe head and torso. Every figure distinguishes between male (blue triangles) and female (red squares) subjects. On average, the

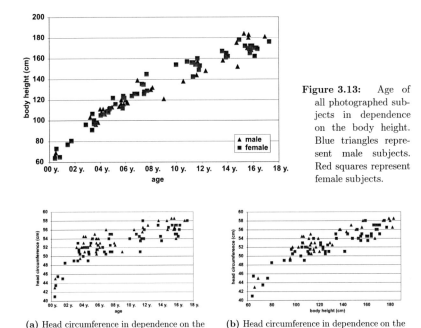

Figure 3.13: Age of all photographed subjects in dependence on the body height. Blue triangles represent male subjects. Red squares represent female subjects.

(a) Head circumference in dependence on the age

(b) Head circumference in dependence on the body height

Figure 3.14: Head circumference of all photographed subjects. Blue triangles represent male subjects. Red squares represent female subjects.

head perimeter and the height of the head for boys are slightly larger than for girls. However, the other values, especially the ear geometries, do not differ significantly between boys and girls. As shown in the figures, the various geometrical data have different characteristics. Some values show a big growth-dependency while others do not. For example the depth of the head (Figure 3.15(c)) seems to reach the final value very early (while children are attending in kindergarten) compared to the height of the head (Figure 3.15(a)). This value increases continuously up to adolescence.

The same phenomenon can be observed in terms of the width and the height of the pinna (Figure 3.15(e) and Figure 3.15(f)).

Figure 3.15(d) shows the distance from the ear canal entrance to the shoulder. This value is very important for the first reflection on the shoulder. This value, however, has a big statistical spread and rises up to the adult age.

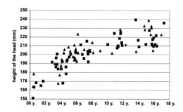

(a) Height of the head in dependence on the age

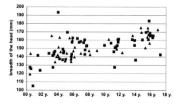

(b) Breadth of the head in dependence on the age

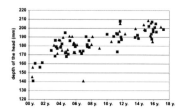

(c) Depth of the head in dependence on the age

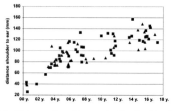

(d) Distance from the ear to the shoulder in dependence on the age

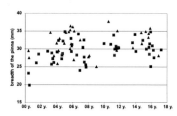

(e) Breadth of the pinna in dependence on the age

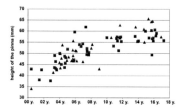

(f) Height of the pinna in dependence on the age

Figure 3.15: Various head parameters of all photographed subjects. Blue triangles represent male subjects. Red squares represent female subjects.

Chapter 4

Anthropometric Data and its Influence on Binaural Cues and HRTFs

This chapter deals with the question how a certain anthropometric parameter can influence the head-related transfer function and binaural cues. During growth, the parameters fluctuate between different values, in some cases drastically, in others only slightly.

First of all, it is settled how much a HRTF from an infant differs from an adult one. Subsequently, it is analyzed whether the down-scaling of (standardized) adult heads to the head circumference of a child yields similar HRTF results as the child itself. On the basis of the data obtained by the photogrammetric system (cf. Section 3.3) a statistical evaluation is carried out. The influence of the most important parameters describing head, pinna and torso are analyzed as far as possible independent from one another.

4.1 Case Study – Infant's HRTF vs. Adult's HRTF

The HRTF of a typical male adult according to the measurements of GENUIT [Gen84] is compared to the HRTF of a six-months-old infant to get an idea of how much a HRTF changes during growth. Figure 4.1 (a) and (b) show the appropriate CAD models. Figure 4.1(c) shows the |HRTF| in decibels (dB) for frontal sound incidence and sound incidence from 60° in the horizontal plane. The abstract model of the adult was chosen, because this model can easily be manipulated and scaled down. Especially when it comes to the incidence from the front, there are tremendous differences between the adult and the child HRTF. At frequencies where the child HRTF has the first dip (at 3.5 kHz) the adult HRTF shows the rise to the second maximum, while the first dip of the adult HRTF is at about 1.5 kHz.

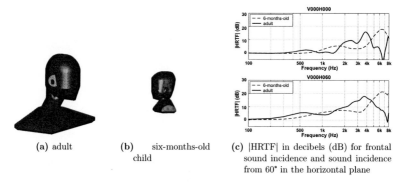

(a) adult   (b)   six-months-old child   (c) |HRTF| in decibels (dB) for frontal sound incidence and sound incidence from 60° in the horizontal plane

Figure 4.1: Head-related transfer functions of an adult and a six-months-old infant

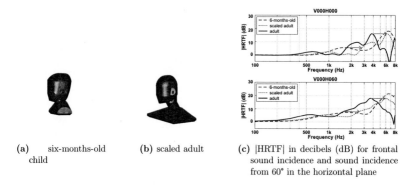

(a)   six-months-old child   (b) scaled adult   (c) |HRTF| in decibels (dB) for frontal sound incidence and sound incidence from 60° in the horizontal plane

Figure 4.2: Head-related transfer functions of an adult, a six-months-old infant and a down-scaled adult head

At first it could be assumed that a down-scaled adult head yields results similar to the child head. However, as shown in Figure 4.1, a scaled adult head with the same head circumference as the child head, yields a frequency shift ([Mid99a, Mid99b]). In this case there are still large differences between the adult and the child head. Now at the point where the child's head reaches the first minimum, the HRTF of the adult has its first maximum.

This initial case study shows that the HRTF of children can not be generated by using simple scaling algorithms. A closer look into the complex system is taken in the following sections to identify which head, pinna and torso dimensions cause which kind of differences.

4.2 Statistics of the Anthropometric Data of the Head and Torso

Geometrical data of 95 children aged from six months to seventeen years has been measured according to Figure 3.11. Therefore the whole database provides a basis for further studies about the entire spectrum, within whom the different parameters can be found.

However, since the collection of the data started in kindergartens, the most subjects are in the age of kindergarten children. 31 of the photographed subjects are kindergarten children. 24 of these children are between four and six years old. The main focus is put on this age group, since the supply with hearing aids begins at the latest in this age - with modern hearing screenings taking place when children are about of six-months old.

Thus an initial statistical evaluation is made on the group of kindergarten age. The deviation from the mean and median values calculated on the basis of the whole kindergarten age group and the children between four and six are less than 0.2 mm. The standard deviations of the kindergarten group are in the same order of magnitude as reported by other authors for adult data [Gen84]. The radii of the back of the head and the angle of the neck are the only exceptions.

The median values are chosen to build a head representing children of kindergarten age. The median is not affected by some very low and high values – in opposition to the mean value. However, there are only slight differences between the median and mean values.

Table 4.1 lists the mean values of the most important parameters describing roughly head, torso and pinna. The values according to the ITU recommendation [ITUP.58, ITUP.57] are listed as well as the adult values according to GENUIT [Gen84] as well as the mean and median values of the kindergarten group.

Table 4.1: Means and medians of anthropometric measures of adults and kindergarteners

measure	adult (ITU nominal)	adult (Genuit [Gen84])	kindergarten mean (4 – 6.5 years)	kindergarten median (4 – 6.5 years)
height of the head	224 mm	261 mm	194 mm	194 mm
breadth of the head	152 mm	177 mm	148 mm	145 mm
depth of the head	191 mm	218 mm	187 mm	186 mm
perimeter of the head	–	62 cm	52 cm	52 cm
distance from ear to shoulder	170 mm	160 mm	87 mm	84 mm
height of the pinna	66 mm	70 mm	54 mm	54 mm
breadth of the pinna	37 mm	35 mm	31 mm	32 mm

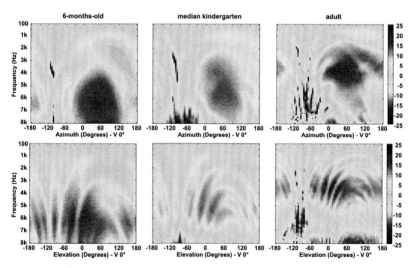

Figure 4.3: Head-related transfer functions of a six-months-old infant, median kindergarten, and adult head in the horizontal and median plane

In the following the HRTFs of the six-months-old child, the median kindergarten child and the adult are calculated and compared to get an impression of the growth dependency HRTFs and Head-Related Impulse Response (HRIR)s. It can be clearly seen that the growth does not cause a frequency shift in the whole horizontal plane. The structure of the HRTF becomes different, too.

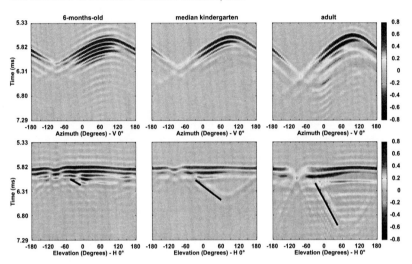

Figure 4.4: Head-related impulse responses of a six-months-old infant, median kindergarten, and adult head in the horizontal and median plane

Figure 4.3 shows the head-related transfer functions of a six-months-old infant, median kindergarten, and adult head in the horizontal and median plane. The colorbar indicate the HRTF-values in decibels. In the median plane (lower row) it seems like the HRTF of a young infant is a zoomed-in part of the adult one.

The head-related impulse responses differ in the structure, too. Figure 4.4 depicts the HRIRs of the three generations in the horizontal and median plane. It can be observed that if the head and the distance between the ear and the shoulder are smaller, different results are obtained. In the median plane a reflection due to the shoulder can be observed (marked by the black line).

The Figures 4.5 and 4.6 show a balloon plot of the HRTFs for all angles of incidences at a frequency of 3000 Hz (Fig. 4.5) and 6000 Hz (Fig. 4.6) for the six-months-old, median kindergarten and adult head. Each figure shows the balloon from different point of views. The coordinate system is the same as introduced in Figure 1.2 and Section 3.1.4, Fig. 3.8. The origin is placed in the center of the head. The HRTFs of the left ear are depicted. This kind of diagram depicts how different the direction

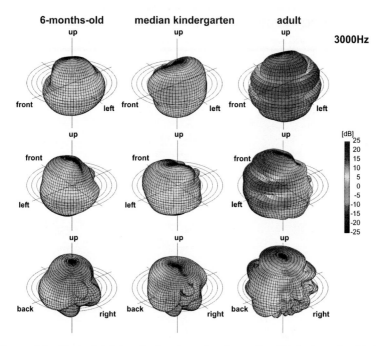

Figure 4.5: Head-related transfer functions of a six-months-old infant, median kinder-garten, and adult head at a frequency of 3000 Hz, viewed from various angles

depended parts are caused by the different geometries of the heads.

Accordingly, the binaural cues "Interaural Time Difference" and "Interaural Level Difference" differ very much, too (see Figure 4.7).

According to other researchers like [ADMT01, AAD01b] and [BD98] (besides many others), significant features of the HRTF and the HRIR can be attributed to anatomical features, as already described in case of the reflection caused by the shoulder in the HRIR in Figure 4.4.

The issue described above shows that it is important and necessary to create artificial heads for children in various age groups. The problem is to figure out, at what age or body height a new artificial head is needed, or in other words, which anthropometric parameter is important for the HRTF and which can be handled with more tolerances.

Therefore, beside focusing on age groups, the growth-dependency can be studied

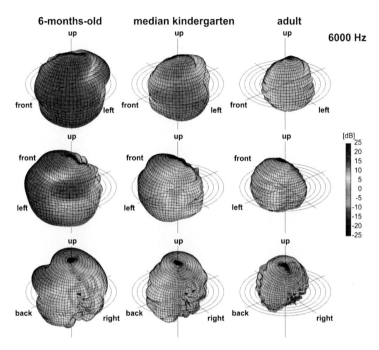

Figure 4.6: Head-related transfer functions of a six-months-old infant, median kindergarten, and adult head at a frequency of 6000 Hz, viewed from various angles

for each parameter. The median and the average value were extrapolated from these datasets for each head measurement. The total population of test subjects is treated as one group because the overall variation in terms of anthropometric data will be analyzed. This makes it possible to analyze which anthropometric parameter grows faster or slower during a certain growth period and which parameter does not show a growth dependency at all. Additional to the parameter study of the whole population, a study with focus on young adults is carried out and described in Section 4.3.5.

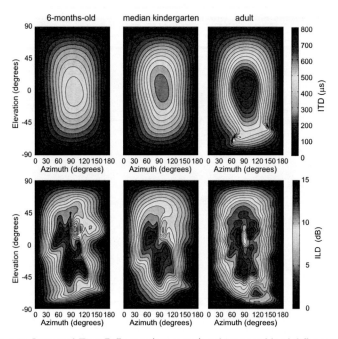

Figure 4.7: Interaural Time Difference (upper row) and interaural level difference (lower row) of a six-months-old infant, median kindergarten, and adult head

4.3 Parameter Variations of the Head and Torso Geometry

In order to investigate the influence of the head and torso geometry, this thesis focuses on six important anthropometric parameters describing roughly a head: the distance from the shoulder to the ear (measured from the *Ear Canal Entrance Point* (EEP)), the frontal vertex (forehead), the back vertex, the breadth of the head, the height of the head, and the distance from the chin (lowest point) to the EEP. With the help of these parameters it is possible to construct a simplified model of the head. Table 4.2 shows these values and their ranges. To evaluate the highest point on the top, or the breadth, for example, the vertex of the radii measured on the heads was taken (as written in detail in Section 3.3). Since this table shows the maximum and minimum values as well, the range of the parameters during growth is visible. The distance

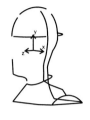

Figure 4.8: Outlines from the 5 % quantile (left), median (middle) and 95 % quantile value heads

from ear to shoulder varies most of all. Additionally, the 5 and 95 % quantiles are calculated.

When comparing these three head sizes, 5 and 95 % quantile and median, with the database, several turned out to match with the actual head sizes of subjects from the above-mentioned database. The 5 % quantile head is equivalent to that of a six-months-old infant, the median head is equivalent to that of a six-year-old child, and the 95 % head to that of a sixteen-year-old teenager. The model heads were all constructed without a pinna as only the influence of the six most important head parameters will be investigated in this part. The influence of the pinna will be discussed in greater detail in Section 4.4.

Table 4.2: Parameter variations of the anthropometric data of the head and torso investigated. Maximum, minimum, mean, median and 5 and 95 % quantiles. Values refer relativly to the center ot the head.

measure	minimum	5 % quantile	mean	median	95 % quantile	maximum
distance ear to shoulder	26 mm	43 mm	99 mm	99 mm	138 mm	157 mm
breadth	53 mm	64 mm	76 mm	76 mm	86 mm	97 mm
chin	44 mm	52 mm	72 mm	72 mm	91 mm	102 mm
top of the head	106 mm	115 mm	131 mm	132 mm	141 mm	150 mm
vertex front	42 mm	70 mm	86 mm	87 mm	98 mm	108 mm
vertex back	69 mm	81 mm	98 mm	98 mm	114 mm	117 mm

Figure 4.8 shows the outlines of the three heads resulting from the 5 and 95 % quantile and median values and their corresponding radii. The complete variation of each parameter can be studied here. The childlike characteristics of the smallest head becomes very clear.

4.3.1 Effects on HRTF

First of all, the median head was built with the help of the database. The purpose of creating a median head was to produce a head based on all values that reflect all dimensions in the middle of the growth. On this basis the geometric measurements can be varied and extended to smaller and larger values. This median head is not intended to represent the "best head" or the like.

First of all the HRTFs are calculated for the horizontal and median plane. Figure 4.9 shows the HRTF magnitudes for the three heads in Figure 4.8. The colorbar indicate the dB-value of the HRTFs.

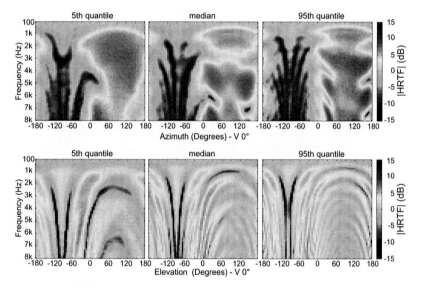

Figure 4.9: HRTFs from the 5th quantile, median and 95th quantile head for the horizontal plane (upper row) and the median plane (lower row)

It can be observed clearly that the three sizes result in completely different HRTFs. For very low frequencies no difference can be observed. This is expectable, since the wavelength is too long to be influenced by the head. However, differences in the magnitude of the HRTF of more than ± 10 dB dB were detected above frequencies of about 1 kHz.

A clear maximum of the HRTF ranging from 3.5 to 5 kHz can be seen for the median head in case of frontal incidence. In this frequency range, the HRTF of the

Figure 4.10: CAD models from the frontal view of the head where only the breadth of the head is changed to the 5 % quantile value (left), the median head (middle) and the head where only the breadth of the head is changed to the 95 % quantile value

5 % quantile head shows a minimum in the HRTF. In the median plane, the arch-shaped notches of the 95 % quantile head are closer together than those of smaller heads.

While using the median head only one parameter was varied at a time to analyze the influence of the individual parameters on the HRTF. For example, Figure 4.10 shows the approach when it comes to the parameter "breadth of head". The breadth was changed here to the 5 % value one time and to the 95 % value the other time. All other parameters were not changed. 12 CAD models, each one differing from the median head in one single parameter only, were created with the help of this approach. In order to highlight the influence of each parameter on the HRTF, the differences of the HRTF magnitude between the median head and the head with the changed parameter are calculated ($|\text{HRTF}(head_1)| - |\text{HRTF}(head_2)|$) for all angles of incidence and frequencies. Examples of these differences are plotted in Figure 4.11.

In the very left column of Figure 4.11 we see the difference in the magnitude of the HRTF for the heads whose parameters are all changed. The difference between a median and the 5 % quantile and the 95 % quantile - heads (according to Fig. 4.8 and 4.9), respectively can clearly be seen. It is of particular importance to notice the influence of the distance between the ear and the shoulder. The differences triggered by this parameter are almost as large as the differences in the results of all 5 and 95 % quantile heads.

Furthermore it can observed that the breadth of the head and the back vertex have an enormous influence as well (the colorbar range from -5 dB to +5 dB). Other parameters such as the chin or the height of the head have little influence on the HRTF.

The assessment of the variations of all head dimensions and of the impact they have on details in the HRTF is a complex challenge. Therefore a difference measure in Eq. (4.1) is used to summarize the results.

$$\text{difference measure HRTF} = \frac{\sum |\text{difference values [dB]}|}{\text{number of values}} \tag{4.1}$$

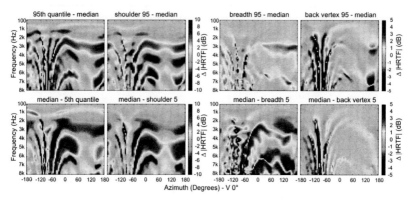

Figure 4.11: Differences in the HRTF caused by changes in the head and torso geometry in dB. Each figure shows the difference between the head with the altered value and the median head in the horizontal plane. In case of the heading "quantile" all parameters are changed.

The differences for certain calculated planes are thus converted into a single-number value. A higher value signifies a high average deviation and a lower value signifies a low average deviation. Since each plane is simulated in steps of 5° and for frequencies from 100 to 8000 kHz, a high number of different angles of incidence are taken into account on each calculated plane. Although a HRTF can have at specific angles different characteristics, in total this measure provides a reliable statement. The differences in the HRTF have been observed at all angles of incidence and no value had outliers.

Table 4.3 shows the difference measure of the HRTFs on different planes. For each plane the difference measure is calculated once for the difference between the head where the parameter was changed to the 95 %-quantile value and the median, and once for the difference between the median head and the head where the parameter was changed to the 5 %-value. The head created by combining all 5 % or 95 % values respectively, is called the "quantile". The parameter "distance between shoulder and ear" has the largest influence on the HRTF. It is interesting that this parameter is also the one that varies most during growth (cf. Table 4.2). When it comes to larger heads (95 % – median, e.g. 16-year-olds – 6-year-olds) the back vertex has a strong influence on the HRTF. The breadth of the head plays a more important role for smaller heads (median – 5 %). All other parameters have only little influence on the HRTF. The results for the horizontal plane and for the plane in 30° elevation are almost the same.

Table 4.3: Average differences [dB] in the HRTF caused by the alteration of the head and torso geometry calculated according to Eq. (4.1) for different planes.

	horizontal plane 95-M	M-5	elevation 30° 95-M	M-5	elevation 60° 95-M	M-5	elevation −30° 95-M	M-5	median plane 95-M	M-5
quantile	3.0	3.1	2.8	3.1	2.2	2.7	2.1	1.9	2.7	2.8
shoulder	2.6	3.1	2.6	3.4	2.1	2.9	1.4	1.4	2.1	2.4
breadth	0.7	2.0	0.8	2.0	0.6	1.3	0.6	1.5	0.6	1.4
height	0.3	0.6	0.3	0.4	0.3	0.4	0.3	0.4	0.2	0.3
frontal v.	0.9	0.8	0.6	0.5	0.4	0.3	0.5	0.4	0.4	0.3
back v.	1.3	0.8	1.0	0.8	1.1	0.4	1.3	0.4	1.4	0.3
chin	1.1	1.0	0.9	0.8	0.8	0.5	0.6	0.9	0.6	0.6

4.3.2 Influence on HRIR

In an analogous manner as the previous section, the influence of each parameter on the HRIR can be investigated. In the HRIR the time structure between the left and right ear is easy to observe. When looking at the three different head sizes, the most outstanding feature is the resulting time delay at various angles of incidence. If the HRIR is plotted in the horizontal plane vs. all angles of incidence, for the 5 %-quantile head there is only a slight time-shift, while for the 95 %-quantile head the delay is very big. This results in a larger ITD. This is mainly caused by the different breadth of the head. Figure 4.12 depicts the influence of the head breadth on the HRIR. The HRIR of the three heads corresponding to Figure 4.10 are displayed in the horizontal plane.

In the median plane the characteristics of the shoulder reflection becomes clear. It is obvious to see, that the shoulder reflection of the smaller head, is direct behind the first reflections. Figure 4.13 shows the HRIR in the median plane for the three heads, where only the distance from the ear to the shoulder is changed. The influence of only that parameter becomes very clear.

The variations of all head dimensions and of the impact they have on details in the HRIR is converted into a single-number value using Equation (4.2).

$$\text{difference measure HRIR} = \frac{\sqrt{\sum \left(\text{difference values}\right)^2}}{\sqrt{\text{number of values}}} \tag{4.2}$$

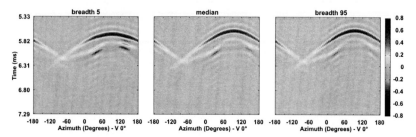

Figure 4.12: Head-related impulse responses of the head with the 5 %-breadth value, the median, and the 95 %-breadth value in the horizontal plane.

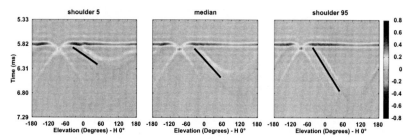

Figure 4.13: Head-related impulse responses of the head with the 5 %-shoulder value, the median, and the 95 %-shoulder value in the median plane.

Corresponding to the single value of the HRTF a higher value signifies a high average deviation and a lower value signifies a low average deviation.

Table 4.4 shows the difference measure of the HRIRs on different planes in the same manner as for the HRTF in the previous section. The values are displayed with a factor of 1000 for reasons of legibility. The absolute values are arbitrary since the impulse responses may be normalized to a certain value, which is not done in this thesis.

Table 4.4: Average differences (factor 1000) in the HRIR caused by the alteration of the head and torso geometry calculated according to Eq. (4.2) for different planes.

	horizontal plane 95-M M-5		elevation 30° 95-M M-5		elevation 60° 95-M M-5		elevation −30° 95-M M-5		median plane 95-M M-5	
quantile	15.4	31.2	15.6	31.2	9.7	15.9	10.1	25.5	10.8	11.4
shoulder	14.0	14.3	13.0	16.4	8.4	11.3	6.6	6.5	9.4	10.3
breadth	7.5	28.8	7.6	25.7	3.8	13.7	7.0	25.9	2.4	4.8
height	0.8	1.4	0.8	1.7	0.8	1.5	0.7	1.0	0.6	0.9
frontal v.	1.8	1.6	2.0	1.4	1.3	0.9	1.4	1.2	1.3	0.9
back v.	4.2	2.1	3.6	2.7	4.7	1.3	4.1	1.5	4.6	1.6
chin	2.5	2.8	2.7	2.0	2.6	1.5	2.0	3.2	2.2	2.1

The parameter breadth of the head has the largest influence on the HRIR for smaller heads (median − 5 %). According to the HRTF results the parameter parameter "distance between shoulder and ear" plays a very important role. The back vertex shows a significant influence on the HRIR, too, while the other parameters (height, frontal vertex and chin) have only little influence on the HRIR.

4.3.3 Influence on ITD

The Interaural Time Difference (ITD) is an important cue for localization. The ITD is computed as the time shift of the maximum of the cross-correlation function of the left and the right ear impulse responses. Since only the impulse responses for the left ear are simulated, the right ear impulse responses can be substituted by the corresponding left ear response. Thus, the corresponding right ear impulse response can be calculated by $incidence(right) = 360° - incidence(left)$, since the modeled heads are symmetric. The impulses were low-pass filtered using a 1500 Hz cut-off frequency before cross-correlation. Figure 4.14 shows the ITD values calculated for the whole sphere. Only half of the sphere is plotted for reasons of symmetry. The colorbar indicate the ITD values from 0 to 800 μs. Each contour curve indicates a step of 50 μs.

The maximum ITD value of the 5 % quantile head is approx. 500 μs, while the median head yields a maximum ITD of approx. 625 μs and the 95 % quantile head about 700 μs. This corresponds to values introduced by [Kuh77].

The differences caused by growth are to be considered very large since the JND of the ITD is approx. 10 μs (e.g. see MILLS [Mil58] or BLAUERT [Bla74] pp. 113-125

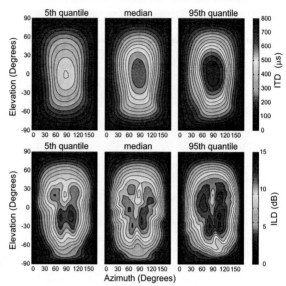

Figure 4.14: ITD and ILDs of the 5 % quantile, median and the 95 % quantile head

(table p. 123)). Once again the breadth of the head, the distance between the ear and the shoulder, and the back vertex are the most influential parameters.

4.3.4 Influence on ILD

The Interaural Level Difference (ILD) is another important cue. The ILD is calculated by subtracting the average energy of the HRTFs of all simulated frequencies (broadband). Figure 4.14 shows the ILD values for the 5 and 95 % quantile and median heads. The colorbar indicate the ILD values from 0 to 15 dB. Each contour curve indicates a step of 1 dB. The JND of the ILD depends on frequency and level and is approx. 1.5 dB (cf. BLAUERT [Bla74] Table, p. 129). Figure 4.15 shows to what extent some parameters differ from the median head. The colorbar indicate the ILD difference values from –2.5 to +2.5 dB. Each contour curve indicates a step of 0.2 dB.

In the ILD difference graphic of the 95 % quantile and median head, the dark red correspond to difference values of more than +2.5 dB. In the case of the difference between median and the head where the breadth is changed to the 5 %-value, the white spots correspond to the ILD difference values of less than –2.5 dB.

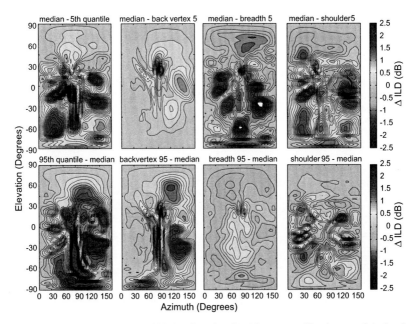

Figure 4.15: Differences in the ILD for all angles of incidence caused by changes of the head and torso geometry. Each figure shows the ILD difference in dB between the head with the altered value and the median head. In case of the heading "quantile" all parameters are changed.

4.3.5 Additional Parameter Variations of Head and Torso Geometry on basis of Young Adults

Additionally, with the knowledge of the parameter variations for a group from six months up to adults, the most important parameters are studied on the variations of 16 young adults (age range from 15 years and 5 months up to 17 years and 4 months) [Fel05]. Here, a statistical evaluation as in the previous sections was made, too.

Table 4.5 shows the maximum, minimum, mean, median and the 5 and 95 % quantiles for the 16 young adults. The variation is, of course, smaller than from the whole database. For the parameter studies, again, CAD models starting from the median data (now equivalent to a 16-years-old) were created. Additionally, the models of according to the 5 % and 95 % quantiles were created. For each of the 13 heads the whole HRTF and HRIR dataset were calculated and the effects of each parameter investigated.

Table 4.5: Maximum, minimum, mean, median and 5 and 95 % quantiles for 16 young adults. Values refer relativly to the center ot the head.

measure	minimum	5 % quantile	mean	median	95 % quantile	maximum
distance ear to shoulder	104 mm	111 mm	128 mm	126 mm	147 mm	149 mm
breadth	71 mm	74 mm	83 mm	83 mm	89 mm	92 mm
chin	71 mm	73 mm	84 mm	85 mm	95 mm	96 mm
top of the head	131 mm	131 mm	137 mm	135 mm	146 mm	150 mm
vertex front	83 mm	83 mm	91 mm	91 mm	101 mm	108 mm
vertex back	97 mm	99 mm	107 mm	106 mm	115 mm	115 mm

Although the differences in the measured data are only a few centimeters in case of the distance of the ear to the shoulder and only 5 to 9 mm in case of the breadth of the head, the differences in the HRTF are clearly visible and exceed ±5 dB. Comparing the results of this small group of young adults with the results of the whole database, the same general trends turned out.

The change in top of the head show almost no influence in the HRTF, although all possible angles of incidence have been analyzed. The frontal vertex seems to be also quite unimportant. The parameter breadth of the head is sensitive at some angles of incidence. The back vertex, however has a big influence at almost all angles of incidence.

4.4 Parameter Variations of the Pinna Geometry

For the experiment described above all CAD models of heads are built without a pinna. In this section the influence of anthropometric pinna parameters will be discussed.

The size, position and orientation of the pinna in relation to the head is measured with the photogrammetric system. A second coordinate system is defined additionally to the head-related one (cf. p. 9), so that various pinnae can be compared despite different head sizes and shapes. The ear canal entrance is defined to be the origin of the pinna coordinate system. The outer pinna is simplified in a plane area. This plane defines the x- and y-axis of the pinna coordinate system, so that the upper and lower radii of the pinna and the edge of the cavum conchae are in one plane. The height and the breadth of the ear, the upper and lower radii and the height, breadth and depth of the cavum conchae are measured from the photographs and stored in a database. The rotation of the ear is calculated with the help of the two coordinate systems. All the pinna data refer to this pinna coordinate system.

Table 4.6: Parameter variations of the anthropometric data of the pinna investigated. Maximum, minimum, mean, median and 5 and 95 % quantiles. Values refer relatively to the pinna coordinate system.

measure	minimum	5 % quantile	mean	median	95 % quantile	maximum
breadth (pinna)	17.9 mm	23.0 mm	28.9 mm	28.7 mm	33.9 mm	37.6 mm
height (pinna) upper vertex $+x$	17.8 mm	23.9 mm	29.2 mm	29.2 mm	34.9 mm	40.0 mm
height (pinna) lower vertex $-x$	14.5 mm	19.2 mm	24.1 mm	24.1 mm	29.6 mm	31.1 mm
breadth (cavum conchae)	12.7 mm	13.3 mm	18.0 mm	18.4 mm	24.3 mm	30.1 mm
height (cavum conchae)	12.2 mm	16.0 mm	21.6 mm	21.5 mm	26.1 mm	28.7 mm
depth (cavum conchae)	4.3 mm	6.3 mm	9.6 mm	9.8 mm	13.8 mm	16.3 mm

All ear parameters except the rotation of the ear show a strong growth dependency. Table 4.6 shows these most important parameters and their range of values. The rotation of the ear is not displayed because of its different unit. The parameter changing most during growth is the ear height.

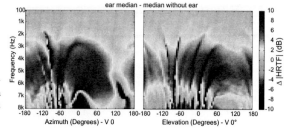

Figure 4.16: HRTF differences of a median head with and without pinna

4.4.1 Effects on HRTF and HRIR

Equivalent to the construction of the head described in Section 4.3.1, a median pinna is constructed based on the median data of the pinna measurements. This median pinna is attached to the median head as described in Section 4.3.1.

The effect of the pinna on the HRTF becomes clear when the median head is compared with and without a pinna. Figure 4.16 shows the influence of the pinna on the magnitude of the HRTF. The difference between the HRTFs of the head with a pinna and the one without a pinna in the horizontal plane (left) and median plane (right) is plotted. The colorbar indicate the differences in the HRTF in [dB]. The ear causes differences in the HRTF of more than ± 10 dB (cf. [ADMT01]).

Figure 4.17 shows the influence of the pinna on the impulse response. The upper row shows the impulse response in the horizontal plane and the lower row shows the impulse response in the media plane. The initial ridge is followed by a sequence of ridges and troughs. The pinna causes a second and third ridge, which are plainly visible between –30° and 120° (cf. [BD98]) in the horizontal plane. In the median plane the influence of the pinna clearly visible by a salient minimum immediately after the first maximum.

Based on the median head with a median pinna, all ear parameters are changed individually to the lower and upper value by means of the database. Additionally the ears where all parameters are changed to the lower and upper values are studied. Figure 4.18 shows the CAD models of the ears whose parameters are all changed to the 5 % quantile, to the median and to the 95 % quantile values.

The effect of all parameters describing the simplified pinna on the HRTF will be studied in this section in the same manner as in Section 4.3.1.

Figure 4.19 shows the magnitude of the HRTFs in the horizontal plane (upper row) and in the median plane (lower row) for the median head without pinna and the ears corresponding to Figure 4.18. For the three different sizes of ears, the differences are easy to detect. Everything except the parameters of the pinna remained

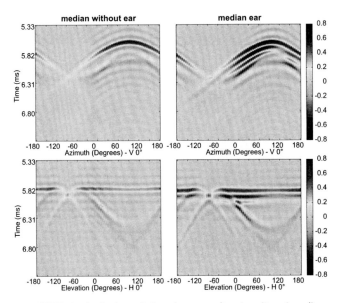

Figure 4.17: HRIRs in the horizontal plane (upper row) and median plane (lower row) of a median head with and without pinna

unchanged. There is an area with higher values in the level of the HRTF (characterized by the white spot). The larger the pinna, the more this spot is shifted towards lower frequencies.

Once again the difference of the HRTFs of an ear with changed value and the median pinna is evaluated to highlight the effect on each parameter on the HRTF. Figure 4.20 shows these differences with regard to several parameters. In each case the parameters are changed to the value according to Table 4.6. In case of the depth which should have been changed to the 5 %-quantile value, the CAD model could not be created as the median head with a median ear turned out to be too large so that the depth with such a value could not be constructed. Therefore the smallest depth available for the CAD model was taken into account. However, this produces an enormous difference, too.

The difference plots show clearly that the pinna causes almost no influence up to 2 kHz. This is, however, as expected since the wavelength is too long to be influenced by the pinna. A significant difference can be seen above frequencies of 3 kHz.

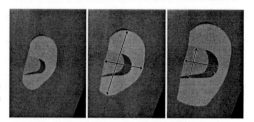

Figure 4.18: CAD models of the 5th, median and 95th quantile pinna

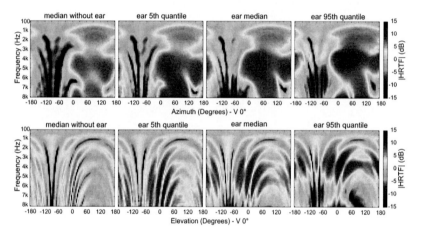

Figure 4.19: HRTFs of the a head with the 5, median and 95th quantile pinna

Once again the difference measure according to Eq. (4.1) is calculated to summarize the effects on the HRTF. Table 4.7 shows the results for the parameter variations of the pinna.

One can easily see that the breadth and the depth of the cavum conchae and the rotation of the ear have the strongest influence on the HRTF. The height and the breadth of the pinna on the other hand turn out to have only little influence even though these parameters vary most of all. Regarding of the depth of the cavum conchae it is expected that this variation causes a direction independent effect in the frequency range of 4 kHz.

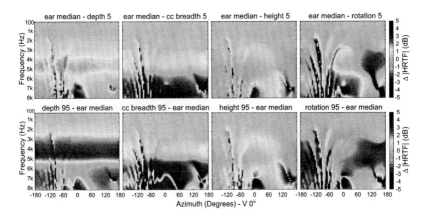

Figure 4.20: Differences in the HRTF caused by changes in the pinna geometry in dB. Each figure shows the difference between the head with the altered value and the median head in the horizontal plane. In case of the heading "quantile" all parameters are changed.

Table 4.7: Average differences [dB] in the HRTF caused by the alteration of the pinna geometry calculated according to Eq. 2 for different planes.

	horizontal plane		elevation 30°		elevation 60°		elevation −30°		median plane	
	95-M	M-5	95-M	M-5	95-M	M-5	95-M	M-5	95-M	M-5
quantile	3.0	2.3	2.9	2.1	3.4	2.1	2.9	2.2	3.2	2.2
rotation	1.4	1.3	1.5	1.0	1.6	1.1	1.0	1.1	1.1	1.0
breadth (pinna)	0.7	0.7	0.5	0.5	0.3	0.3	0.5	0.6	0.5	0.5
height (pinna)	0.7	0.5	0.4	0.3	0.3	0.2	0.6	0.5	0.5	0.4
height (cavum conchae)	0.7	1.0	0.6	1.0	0.6	1.0	0.8	1.1	0.7	1.1
breadth (cavum conchae)	1.8	1.5	1.8	1.3	2.8	1.3	1.7	1.5	2.0	1.4
depth (cavum conchae)	1.7	0.8	1.6	0.7	1.5	0.7	1.7	0.8	1.6	0.8

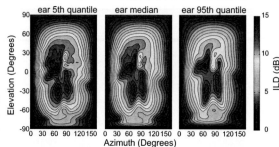

Figure 4.21: ILDs of the head with the 5th, median, and 95th quantile pinnae

4.4.2 Influence on ITD

The different shapes of the ear have, as expected, very little influence on the Interaural Time Difference. The ITD plays a more important role at lower frequencies (below 1500 Hz, as already described in Section 1.2.2). The influence of the pinna, however, begins at frequencies above 2–3 kHz. The influence on the hearing sensation at low frequencies is probably almost non-existent, since the barely noticeable difference is bigger than the difference caused by the varying geometries of the ear. Therefore the effect on the ITD will not be discussed here in detail.

4.4.3 Influence on ILD

In contrast to the Interaural Time Difference, the Interaural Level Difference caused by different shapes of the ear, can be observed very clearly. A plot is chosen, where all the angles of incidence are plotted, to highlight the differences here. The colorbar indicate the differences in the ILD values at each angle of incidence in steps of 5°. Figure 4.21 shows the ILD values for the 5 % quantile-ear, the median ear, and the 95 % quantile-ear corresponding to Fig. 4.18.

Figure 4.22 shows the differences in the ILD for the different parameters changed in comparison to the median ear. Even these minor changes cause differences in the ILD of more than ± 2.5 dB.

4.5 Conclusions

The first conclusion of this chapter is that it is not sufficient to approximate children head and torso models with a scaled adult geometry. The scaling to the same head perimeter yield results that do not represent the anatomy of children appropriately. This leads to different HRTFs and binaural cues, so that it can be concluded that the

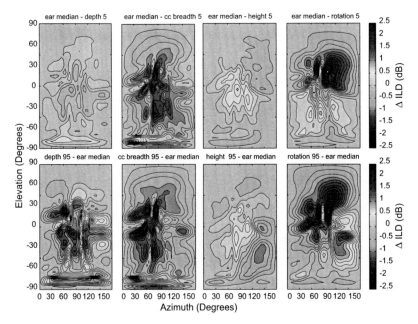

Figure 4.22: Differences in the ILD for all angles of incidence caused by changes of the pinna geometry. Each figure shows the ILD difference in dB between the head with the altered value and the median head. In case of the heading "quantile" all parameters are changed.

binaural hearing and localization processes based on these different cues are inappropriate. The differences between a scaled adult and a child cannot be compared to the individual differences that occur among adults. Regarding individual differences of adults HRTFs, the individual characteristics are usually similar to a typical "mean" HRTF, with more or less deviations. In case of the children's HRTFs, the typical characteristics differ tremendously compared to those of adults.

Taking a closer look at the growth dependency of the parameters describing head and torso, it turns out that some parameters show a enormous growth dependency while others do not. Additionally some parameters grow linearly while others reach a saturation at an age of six to seven.

The relation between body height and age is not linear (Fig. 3.13), which is reflected in other values, too. The head circumference grows more linearly with the body height than with the age. However, other parameters like the breadth of the pinna show a wide spread, and a strong dependence on the age or on the height could not be detected.

Table 4.8: Classification of anthropometric parameters of the head regarding the impact on the HRTF and binaural cues

measure	less important	very important
1. distance ear to shoulder		X
2. breadth		X
3. vertex back		X
4. chin	X	
5. vertex front	X	
6. top of the head	X	

The influence of the parameters described above, are tested independently from each other as far as possible. This kind of procedure is only an attempt to categorize the influence of the growth, however, it is not an orthogonal parameter system. By varying one parameter at a time only an indication is determined. The results, however, are clear and reasonable, which makes it possible to use them as a basis for further studies or applications. Thereby the 5 % and 95 %-quantiles are calculated of each parameter and compared to the median. Thus, this study is independent on the age or the body height, since the whole variation of the values are tested to figure out, which parameter cause a large difference. Based on the knowledge that one parameter has a great impact on the binaural cues, the focus needs to be put on this parameter when it comes to creating age groups or suggesting proper values for artificial heads for children.

In the first experiment (Section 4.3) it turned out that the most important parameter with regard to differences in the HRTF and binaural cues, is the distance between the ear and the shoulder. It is shown that the breadth of the head and the back vertex have an immense influence as well. Only a very slight influence may be attributed to the height of the head, the chin, and the frontal vertex.

Table 4.8 lists the parameters according to their impact on the binaural cues and HRTFs. An attempt was made to sort the data and divide them into more important and less important parameters.

In the second experiment (Section 4.4) the parameter varying most of all during growth, i.e. the height of the ear, turned out to have only very little influence on the HRTF and the computed ILD. The greatest influence has to be attributed to the depth and the breadth of the cavum conchae. Table 4.9 lists the parameters of the pinna in the order of their impact on the binaural cues and HRTFs.

Table 4.9: Classification of anthropometric parameters of the pinna regarding the impact on the HRTF and binaural cues

measure	less important	very important
1. breadth (cavum conchae)		X
2. depth (cavum conchae)		X
3. rotation		X
4. height (cavum conchae)	X	
5. breadth (pinna)	X	
6. height (pinna)	X	

The most important anthropometric data are thus available and can be used for future experiments featuring artificial heads and dealing with the special requirements with regard to children.

It will certainly be useful to take a look at the growth dependency of the most important parameters to create new artificial heads for children.

The distance between the ear and the shoulder ranges between 26 mm and 157 mm. The 26 mm distance occurs since babies cannot hold their head upright or sit up straight by themselves. So the head slumps and the distance between head and shoulder is very short The median value of this parameter is reached at the age of approx. five years, with a small individual spread. Thus, the period from birth to five years is very important, while the rest of the growth is more slowly and with a bigger spread in adolescence (during 13 years). The situation is slightly different for the breadth of the head. This parameter varies between 53 mm and 97 mm (for one half of the head). The median value (76 mm) is reached at the age of approx. nine to 10 years. Thus, this parameter increases more linearly until adulthood is reached.

The most important parameters of the ear are the depth and the breadth of the cavum conchae. In case of the ear, it is difficult to deal with these parameters. The parameters which show a very clear dependency on the growth, is the height of the ear and the breadth of the ear. These parameters, however, show only little influence on the binaural cues. The depth and the breadth of the cavum conchae have a large individual spread. The breadth shows a slight growth dependency and the median value of the whole range is reached at the age of approx. five to six years. However, the individual spread is of the depth of the cavum conchae is very large, thus, in this case the values do not refer to a certain age group.

The pinna is responsible for differences in the binaural cues, but not to such an extent as the head and torso dimensions which cause bigger differences. Thus, when the attempt is made to suggest artificial heads for different age groups, the focus can put on the results of (Section 4.3).

Comparing the data obtained for the oldest age group with the values suggested in [IEC60959], respectively [ITUP.58, ITUP.57] and the data obtained by GENUIT [Gen84] some interesting points occurred.

The values of the pinna dimensions obtained for this thesis are in a good concordance with the normative data, in contrast to the dimensions of the head and torso. The values obtained for this thesis show a good concordance to the values measured by GENUIT, but differ tremendously from the standardized values. All parameters proposed in the international standard are too small and equal more or less a seven to eight years old student. The only exception is the distance between the ear and the shoulder. This parameter is larger in the standard compared to the data obtained for this thesis. Thus, another important task will be to re-evaluate the standards for artificial heads for adults.

Part II
Ear Canal

Chapter 5

Determination of Ear Canal Impedances

In Part I all geometrical features of the ear from the outer ear and torso to the ear canal entrance are discussed. Due to these features the head-related transfer functions and their cues are directionally dependent (as already shown in Fig. 1, p. 2). In this part the directional independent part will be discussed. Starting from the ear canal entrance, the ear canal geometry and its impedance will be analyzed with regard to the growth dependency.

Since the knowledge of the ear canal impedance plays an important role, for example when it comes to fitting hearing aids (cf. Chapter 1.3), or when the function of the middle ear is investigated many studies have been carried out in this field. Already back in 1956 MORTON AND JONES [MJ56] measured the acoustical impedance. Numerous studies were carried out by HUDDE from 1983 to 1999. Different measurement techniques were discussed [Hud83], but in most cases the focus was put on the eardrum impedance of adults.

RABINOWITZ [Rab81] measured the acoustic immittance in the human ear canal for frequencies ranging from 62 Hz to 4 kHz. STINSON [Sti90] collected data of the acoustic energy reflection coefficients at the eardrum for frequencies between 3 and 13 kHz for 20 ears. CIRIC AND HAMMERSHØI [CH06, CH07] used an impedance tube to measure the acoustic impedances of the ear canal. This method leads to very reliable results, but it would be impossible to apply this method while working with children, since one cannot use the apparatus with children easily.

Only a few studies addressed the issue of ear canal properties of children: KRUGER AND RUBEN [KR87] measured the acoustic properties of infants and toddlers. The subjects were between 0 and 40 months old. A probe was inserted 2 mm in front of the eardrum. The results show that the resonance frequency is a function of age. The resonance frequency of newborn children is at about 7 kHz while the one of 20 to 37 months-old children is about 2.7 kHz, which is already similar to an average adult.

PFEIL [Pfe79] chose an indirect method to obtain ear canal parameters. The ear canals of 31 children, who were between two and ten years old, were filled with a cold curing synthetic material up to the eardrum while they were undergoing surgeries under general anesthetic. The negatives of the ear canal were measured regarding their geometrical features. The average volume turned out to be $0.6\,\text{cm}^3$.

KEEFE ET AL. [KBAB93] made a comprehensive study of ear canal impedances and reflection coefficients of infants and adults. Measurements of an adult group and groups of infants who were 1, 3, 6, 12, and 24 months old, were carried out over a frequency range of 125 to 10700 Hz. KEEFE ET AL. [KBAB93] underlined the growth dependency of the measured parameters. When it comes to adults the ear canal impedances ramp down with 6 dB/octave until the minimum of the impedance at 6 kHz. Up to higher frequencies the impedance rises with 6 dB/octave. The impedance functions of the 1 to 24 months-old children do run not in that straight line. Their functions, however, increasingly resemble the adults' functions the older the children get.

HELLE [Hel83] analyzed different procedures to determine the effective gain of hearing aids, when measured at the IEC 60711 coupler compared to the child's ear. He also postulated correction factors for this kind of measurement.

However, detailed data about the ear canal impedances, which refer to the ear canal entrance plane located at the intersection between cavum conchae and ear canal for adults and children are very rare. In this thesis the ear canal impedances are measured and simulated on adults as well as on children. Both methods are comparable since they refer to exactly the same reference plane and the same definition of the ear canal impedance.

5.1 Measurement

The measurement of ear canal impedances is carried out using a method introduced by LODWIG AND HUDDE [LH95]. They applied this method to ears of adults. Now, for the first time, the viability of this approach is tested on children. This method features a one-microphone impedance probe using the principle of the calibrated source. This arrangement includes one loudspeaker mounted on a tube with an adapter (see Figure 5.1). The loudspeaker injects the sound into the tube. Lateral on the tube there is a small hole for the microphone. This microphone measures the sound pressure $\underline{p}$, which depends on the attached load.

This measurement technique is based on the description of the loudspeaker with regard to a reference measurement plane M by an ideal source with a series impedance $(\underline{p}_0, \underline{Z}_0)$. Once these parameters are known, the source is fully described and the impedance of the ear canal can be measured by only one measurement of the sound pressure on the reference plane M. Later on the impedance at the measurement position is transformed to the ear canal entrance plane.

At least two measurements with known terminators are needed to determine the source parameters $\underline{p}_0$ and $\underline{Z}_0$. The simplest approach is the use of terminators in two different lengths, which are easily described mathematically.

The advantage of this measuring method is that the measurement probe is very small, which makes it possible to use it while working with children. Fur-

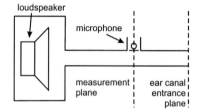

Figure 5.1: Model of the impedance probe with loudspeaker and microphone.

thermore it is easy to handle the calibration. Figure 5.2(a) shows the technical drawing of the impedance probe. The loudspeaker which is used, is an aluminum calotte loudspeaker. The dimensions of the housing are selected in a way that the loudspeaker operates against a small amount of air. This causes even at low frequencies a high sound pressure level in the connecting tube. Figure 5.3 shows the open impedance probe, the loudspeaker and the microphone.

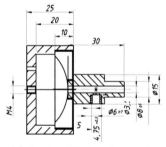

(a) Drawing of the impedance probe

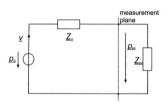

(b) Equivalent circuit diagram with source sound pressure $\underline{p}_0$

Figure 5.2: Impedance probe: Drawing and equivalent circuit diagrams

(a) Open impedance probe: Aluminum calotte and connecting tube

(b) Connecting tube with bore hole for the microphone (KE-4 capsule)

(c) Closed impedance probe

Figure 5.3: Impedance probe

A Sennheiser KE4-211-2 capsule (electret condenser microphone) is placed lateral onto the tube into a small hole. The distance between loudspeaker and microphone was selected in a way so that it would be large enough that higher order modes are decayed. Additionally, the part of the tube between loudspeaker and microphone is damped using damping wool to avoid high resonances in the tube. The probe used for this thesis has an inner diameter of 3.5 mm and an outer diameter of 5 mm. The tube is 25 mm long. Additional test were carried out in FEICKERT [Fei07] with smaller tubes for very young children.

A sweep (order 16 = 32768 samples, sampling rate = 44100 Hz) is taken as the excitation signal. A low-shelf-filter is integrated in the excitation signal to improve the signal-to-noise ratio.

Figure 5.2(b,c) shows the equivalent circuit diagrams of the impedance probe. In the following the diagram describing the source using the source sound pressure $\underline{p}_0$ and series impedance $\underline{Z}_0$ is used. Two different terminators need to be measured to determine the source parameters $\underline{p}_0$ and $\underline{Z}_0$. Two rigid aluminum terminators (cylindrical tubes) of different lengths were created for this thesis.

Using the transmission line equations

$$\underline{p}\,(l) = \underline{p}\,(0)\cos\,(kl) + \mathrm{j}\rho_0 c\underline{v}\,(0)\sin\,(kl) \tag{5.1}$$

and

$$\underline{v}\,(l) = \mathrm{j}\frac{\underline{p}\,(0)}{\rho_0 c}\sin\,(kl) + \underline{v}\,(0)\cos\,(kl) \tag{5.2}$$

yields the impedance version transmission line equation

$$\underline{Z}\,(l) = \frac{\underline{Z}\,(0) + \mathrm{j}\rho_0 c\tan\,(kl)}{\mathrm{j}\underline{Z}\,(0)\tan\,(kl) + \rho_0 c} \tag{5.3}$$

where $\underline{Z}\,(l)$ the impedance 'seen' at the input of a loss free transmission line of length l, terminated with an impedance $\underline{Z}\,(0)$

with

$$k = \frac{2\pi f}{c} = \frac{\omega}{c}. \tag{5.4}$$

Applied on the impedance probe, where $\underline{Z}(0)$ equals the demanded ear canal impedance, is $\underline{Z}(0)$ given by

$$\underline{Z}(0) = \frac{\mathrm{j}\,(\rho_0 c)^2 \tan(kl) - \underline{Z}(l)\,\rho_0 c}{\mathrm{j}\underline{Z}(l)\tan(kl) - \rho_0 c}. \tag{5.5}$$

With the rigid terminators for the calibration, we have $\underline{Z}(0) = \infty$, thus $\underline{Z}(l)$ is given by

$$\underline{Z}(l) = \mathrm{j}\rho_0 c \cot(kl). \tag{5.6}$$

According to the equivalent circuit Fig. 5.2(b)

$$\underline{p}_0 = \underline{v}\,(\underline{Z}_0 + \underline{Z}_{\mathrm{M}}) \tag{5.7}$$

and

$$\frac{\underline{p}_0}{\underline{p}_{\mathrm{M}}} = \frac{\underline{Z}_0 + \underline{Z}_{\mathrm{M}}}{\underline{Z}_{\mathrm{M}}}. \tag{5.8}$$

If the probe is applied at two different terminations, A and B, $\underline{Z}_0$ is given by

$$\underline{Z}_0 = \frac{\underline{p}_{\text{measurement B}} - \underline{p}_{\text{measurement A}}}{\frac{\underline{p}_{\text{measurement A}}}{\underline{Z}(l_{\mathrm{A}})} - \frac{\underline{p}_{\text{measurement B}}}{\underline{Z}(l_{\mathrm{B}})}} \tag{5.9}$$

with

$$\underline{Z}(l_{\mathrm{A,B}}) = \mathrm{j}\rho_0 c \cot(kl_{\mathrm{A,B}}). \tag{5.10}$$

The operating range of the probe is between 100 Hz and 8.5 kHz. The lower limit can be put down to the loudspeaker, since creating a sufficient sound pressure level at low frequencies in the thin connecting tube is rather complicated. The upper limit can be put down to the calibration with the two different rigid terminators. The rigid termination yields a pole on the basis of the cotangent-function.

Using these equations the impedance probe can be fully described and the impedance can be calculated with just one measurement of the sound pressure. Finally the results need to be transformed from the measurement plane to the ear canal entrance plane. The impedance needs to be multiplied with the ratio of the surface area of the tube to the surface area of the ear canal entrance where the probe is attached to.

There are different possibilities to connect the probe and the ear canal. Ear-plugs and tympanometry olives (Fig. 5.4(a, b)) turned out to be insufficient [Fei07]. It is very complicated to handle the foam of the ear-plugs. The probe needs to be fixed

(a) ear protection plug   (b) Tympanometry olives   (c) Individual ear mold (oto-plastic)

Figure 5.4: Coupling devices

into the ear-plug, but they do not seal enough, neither the probe nor the ear canal entrance. This causes problems regarding the repeatability of the measurement. The same holds true for tympanometry olives. Individual formed otoplastics turned out to be the best option. Therefore an impression of the ear is taken, which is usually used to create an ear mold for hearing aids. This impression is used to fit the probe in the ear.

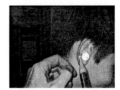

(a) Inspection of the ear canal   (b) Creation of an individual ear mold (1)   (c) Creation of an individual ear mold (2)

Figure 5.5: Manufacturing of the ear mold

The individual ear mold is created with the help of a hearing aid consultant. First, he checks whether the ear canal is healthy and clean (Fig. 5.5(a)) and studies the curvature of the ear canal. Then he places a small wad of cotton wool into the ear canal in front of the eardrum. A piece of thread is tied to the wad to retrieve it from the ear canal. The ear canal and the complete cavum conchae are filled with a two-component cold-curing composition (Fig. 5.5(b)). A pin resembling the size of the impedance probe is placed in the direction of the ear canal and is fixed with the ear mold before curing. The ear mold is cured and can be taken off five minutes later. Figure 5.6(a) shows an individual ear mold directly after taking the ear mold out off the ear canal.

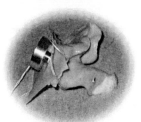

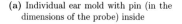

(a) Individual ear mold with pin (in the   (b) Impedance probe ready to measure at
dimensions of the probe) inside   an individual

Figure 5.6: Individual ear mold

The individual impression allows a very exact definition of the ear canal entrance plane since the transition from cavum conchae to the ear canal entrance is clearly visible. The pin is taken out of the ear mold, and the ear mold is cut to the desired length using a scalpel. Then the impedance probe can be placed into the hole of the pin. Figure 5.6(b) shows the impedance probe with an individual ear model ready for the measurement. The ear mold is covered with a thin film of oil to seal the ear mold to the cavum conchae for the measurement.

The measurement procedure is evaluated and the repeatability of the measurement is tested on adults prior to the measurements on children.

The soft ear plugs and tympanometry olives show how the ear canal impedance is distorted, when the impedances probe does not seal the ear canal entrance properly. In this case the impedance or the measured sound pressure equals a measurement with the open impedance probe (with no coupling to the ear canal) at very low frequencies. In this case the impedance is at low frequencies very low.

This might also occur when an individual ear mold does not seal the cavum conchae and the ear canal entrance sufficiently. However, if for each subject the measurement is repeated several times, the flawed measurements can be rejected.

The influence of the quality and repeatability of different individual ear molds is tested on one subject. The hearing aid consultant created an individual ear mold three times. The factor, which may cause differences is, in this case, the placing of the pin during the curing of the ear mold. However, the repeatability of the measurement is very satisfying. Figure 5.7 (a) shows the sound pressure levels and phase responses for the three individual otoplastics.

The maximum deviation is about 2 dB between 100 Hz and 6 kHz.

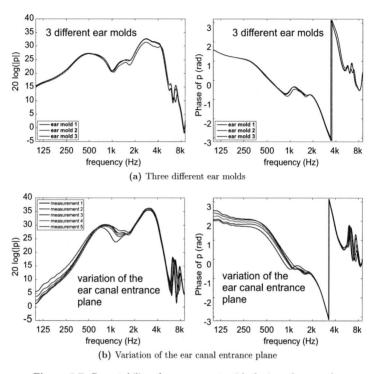

(a) Three different ear molds

(b) Variation of the ear canal entrance plane

Figure 5.7: Repeatability of measurements with the impedance probe

The choice of the ear canal entrance plane has a certain influence. The latter is tested on one subject. The individual ear mold was shortened in 1 mm steps. Thus, at first the ear mold extended 5 mm into the ear canal. Figure 5.7 (b) shows the sound pressure levels and phase responses of the five measurements. Most differences caused by the choice of the ear canal entrance plane occur below 2.5 kHz. However, this measurement deals also with the measurement repeatability itself. Since for each measurement the probe is replaced in the ear. Thus, it is difficult to retrace the reason for the measurement deviations. However, it can be observed that the choice of the entrance plane is rather difficult. However, when using an individual impression, it is comparatively easy to figure out how to define the ear canal entrance plane.

Each measurement it is tested to make sure that the level of the measurement sweep is low enough so that the stapedius reflex does not occur. The stapedius and tensor tympani muscles of the ossicles contract when the sweep is presented with a high-intensity. This should be avoided in the measurement.

5.2   Simulation

Previous studies addressing the simulation of ear canals mainly dealt with the auditory ossicles (e.g. [GS02], [KWK02]). VALLEJO ET AL. [VDH+06] simulated the impedance of one ear canal at the ear canal entrance and compared the results with other authors. STINSON AND DAIGLE [SD05] applied the *Boundary Element Method* (BEM) to study the sound field in the human ear canal.

In this thesis the *Finite Element Method* (FEM) is applied to calculate the ear canal impedance. CAD models of the ear canals are reconstructed for the simulation on the basis of CT scans (computed tomography) of the petrous bone. The construction of the CAD models is the major challenge when it comes to simulating ear canal impedances of children. Whether the reconstruction is finely detailed or not depends on the resolution of the CT slides. In general the simulation works for adults as well as for very young children.

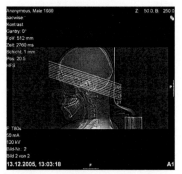   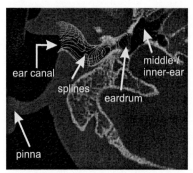

(a) Overview of the cross sectional planes of   (b) Close up of one CT slide. The splines of
a CT-scan of the petrous bone   other slides on top of the other

Figure 5.8: Generation of the CAD models of the ear canals using CT-scans

On each slide the transition between air and tissue is clearly visible. The eardrum is visible as a thin membrane. Thus, the contours of the ear canal and the eardrum can be extracted on each slide. Figure 5.8(a) shows an overview of the cross sectional planes of a CT scan. Every slide is loaded in the program MicroStation (by BENTLEY with the photogrammetry-plugin PHIDIAS [BS97]. The program automatically shifts each slide in the correct distance and angle. Figure 5.8(b) shows a screenshot to illustrate this approach. Consecutively, the CT-scans are loaded and the contours (green) are extracted. The position of the eardrum is marked red. At the ear canal entrance all splines are cut to define the reference plane in the transition between cavum conchae and ear canal. A closed volume model can be generated on the basis of these layered splines (see Fig. 6.3). The volume is discretized using tetrahedral-elements with a length of 1 mm to 3.1 mm (maximum) (corresponding to $d_{max} = \lambda/6$). A total of 25 CT-scans was available for the simulation for this thesis. The test group consisted of 12 male and 13 female subjects aged between three weeks and 20 years.

Using the Finite Element Method the ear canal impedance can be derived up to frequencies of 16 kHz. In the finite elements of each modeled ear canal, the energy formulation of the harmonic field equations is used. This is generally known as HAMILTON'S PRINCIPLE of minimum energy, which is the basis for the finite element method [Zie77]. For each element, the relation between the forces and the displacements is introduced by using the variational approach, which is used to identify the field quantities for minimum energy for each element.

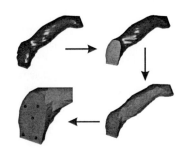

Figure 5.9: Process of CAD modeling of the ear canals

All elements' entries are combined into a matrices for stiffness, mass and damping. A matrix equation is established including the matrices and the source contributions and boundary conditions, which is solved to obtain the sound pressures at certain nodes (detailed information can be found in [Joh87]). The solver *SoundSolve* used for this thesis was developed at the Institute of Technical Acoustics (RWTH Aachen University) (cf. BARTSCH [Bar03], FRANCK [Fra03]). The FEM simulation in this thesis uses the direct solution to determine the eigenvalues of the matrix. The matrix equation is used without further subspace conditions, because of the small models.

The ear canal entrance plane is used as a well-defined interface, but also as an excitation area for the FEM. Since the ear canal is sealed with this plane any reflections caused by this plane have to be avoided. Therefore an impedance of $Z_0 = \rho_0 c_0$

(characteristic impedance of air) is used on the entrance plane. The excitation on the entrance plane is made with a plane wave with a certain particle velocity.

Furthermore, the acoustic properties of the other surfaces (ear canal wall and eardrum) need to be defined. As already described in Section 2.2 it can be concluded that the ear canal walls behave rigid.

It is, however, not easy to determine the acoustic properties of the eardrum. It is hence necessary to model every detail accurately for a correct simulation. There is a myriad of studies dealing with the acoustic properties of the eardrum, especially the eardrum impedance.

From 1956 to 1977, MORTON AND JONES [MJ56], ZWISLOCKI [Zwi70], BLAUERT AND PLATTE [BP76], and MEHRGARDT AND MELLERT [MM77] collected different data of human adults. Additionally, measurements of cadavers were carried out by ONCHI [Onc61] and FISCHLER ET AL. [FFSR67]. However, a large discrepancy between the studies exists. This is caused on the one hand by different measurement techniques and subjects (since ears of dead people are not comparable to ears of living subjects) and on the other hand by the assumption that an ear canal is a tube of constant cross section.

HUDDE [Hud83] showed that the eardrum impedance depends on the variation in the cross-sectional area along the ear canal axis. Additionally the coupling of a measurement probe to the ear is a difficult matter. LETENS AND HUDDE [LH85] studied the eardrum impedance with the help of ear molds which seal the ear canal almost completely.

More recent studies do not confirm these studies and results. HUDDE explains in [HE98a], that in-vivo measurements are only valid up to frequencies of 2 kHz. The latest results are published in [HE98a, HE98b, HE98c]. The results are based upon the measurements of ear canals of corpses. As soon as possible the ear canal were measured (no more than 2 days between death and measurement) and stored in sodium chloride solution.

HUDDE determined an eardrum impedance based on these measurements. This impedance is used in this thesis additionally to the rigid definition of the eardrum. It is still uncertain whether these values can be used for children and infants as well or not.

The results of the Finite Element Method reflect the sound pressure at a definite point (cf. Figure 5.9). With the boundary condition of the ear canal entrance plane ($Z_0 = \rho_0 c_0$), the intended ear canal impedance equals

$$\underline{Z}_e = \frac{\underline{p}}{\underline{v}_2} \tag{5.11}$$

with

$$\underline{v}_2 = \underline{v}_0 \cdot \frac{\underline{Z}_0}{\underline{Z}_0 + \underline{Z}_e} \tag{5.12}$$

($\underline{v}_0$ = particle velocity of excitation).

Thus, the ear canal impedance can be calculated according to

$$\underline{Z}_e = \frac{p}{\underline{v}_0} \cdot \frac{1}{\left(1 - \frac{p}{\underline{v}_0 \underline{Z}_0}\right)} \tag{5.13}$$

with $\underline{Z}_0 = \rho_0 c_0 = 414 \ \mathrm{Ns/m^3}$.

Figure 5.10(b) shows the magnitude of a typical correction function $\underline{Z}_{\mathrm{corr}}$

$$\underline{Z}_{\mathrm{corr}} = \frac{1}{\left(1 - \frac{p}{\underline{v}_0 \underline{Z}_0}\right)}. \tag{5.14}$$

The ear canal impedances and corresponding figures in this thesis are normalized to the characteristic impedance of air according to

$$\underline{Z}'_e = 20 \log \left(\frac{|\underline{Z}_e|}{Z_0}\right). \tag{5.15}$$

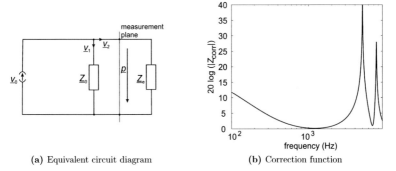

(a) Equivalent circuit diagram   (b) Correction function

Figure 5.10: Calculation of the ear canal impedance based on the sound pressure of the simulation

5.3   Comparison of Simulation and Measurement

The impedance of a well-defined test volume is once simulated using FEM and once measured with the measurement probe to compare the two approaches used to determine the ear canal impedance.

The test volume is a compact aluminum cylinder with an inner diameter of 8 mm. The cylinder is 35 mm long.

After mounting the impedance probe the cylinder is still 25 mm long. This arrangement is similar to the real position on a human ear. The length of 25 mm corresponds to a not yet full-grown ear canal. Placing the probe on the test volume causes a cross-sectional jump which is in accordance to the real ear situation with the probe mounted and connected to the ear canal. Figure 5.11 shows a CAD model and a technical drawing of the test volume.

Figure 5.11: Test volume

The measurement and simulation results are discussed in the following.

Figure 5.12 shows in (a) the magnitude and the phase (b) of the impedance which were observed during simulation (red) and measurement (blue). The frequency and phase responses of the simulation and the measurement show a good concordance. The results of the measurement, however, show a lower quality factor than the simulation. Thus, the phase-jumps in the phase response of the measurement are not so steep.

This might be due to damping, edge effects, or incorrect boundary conditions during the simulation. The simulation is carried out with ideal boundary conditions, which should fit into the case of an aluminum cylinder. But on the other hand the measurement procedure might be not as exact.

When looking at the real (Fig. 5.12(c)) and imaginary part (Fig. 5.12(d)) the differences between simulation and measurement become clearer. The imaginary part shows a very good concordance. The real part, however, shows significant deviation.

If a little damping is applied on the rear panel of the test volume (similar to the eardrum) in the simulation, the two approaches become more similar when looking at the magnitude and phase response of the impedance (Figure 5.13).

Because of the damping during the simulation, the quality factor changes, too. When applying a damping of 10 % during the simulation the quality factor of measurement and simulation is almost the same. However, the phase response shows

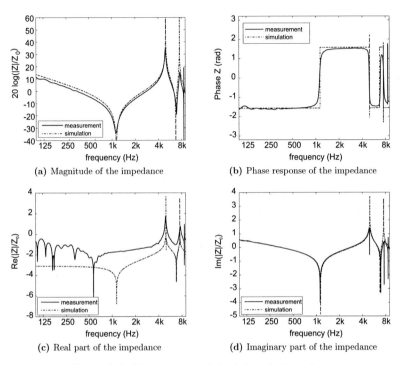

(a) Magnitude of the impedance   (b) Phase response of the impedance

(c) Real part of the impedance   (d) Imaginary part of the impedance

Figure 5.12: Measurement and simulation of a test volume

deviations for low frequencies. This might be due to the assumption that the damping is constant over all frequencies. While calculating the real and imaginary part, differences can still be obtained in the same order without damping as well. This shows that this approach does not lead to satisfying results. It still needs to be determined what kind of physical conditions yield these differences. Additionally, no damping is assumed in this thesis, since the relevant literature provides no data about the correct value of the damping in the ear canal. However, the impedances (magnitude and phase response) show a good concordance.

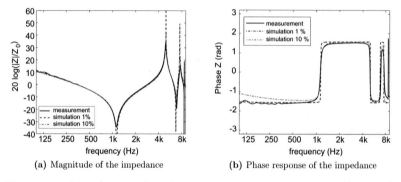

(a) Magnitude of the impedance   (b) Phase response of the impedance

Figure 5.13: Measurement and simulation of a test volume: Damping of 1 % and 10 % added in the simulation

Chapter 6

Anthropometric Data and its Influence on Ear Canal Impedances

6.1 Ear Canal Parameters

When creating the CAD models for the simulation, the models can be used for a data analysis of the geometrical features of the ear canals. The ear canals of 25 subjects between three weeks and 20.5 years of age (Fig. 6.3(a)) are modeled with the help of CT-scans (cf. Section 5.2). The parameters ear canal volume, length and the surface area of the eardrum are calculated and discussed.

Ear Canal Volume

Many different values of the ear canal volume can be found in the relevant literature. BRUNNER [BN01] published a volume of $1.5\,cm^3$, STINSON ET AL. [SL89] determined values between $0.91\,cm^3$ and $1.725\,cm^3$ for subjects with an average age of 55 years for the male and 65 years for the female subjects. PFEIL [Pfe79] determined an average value of $0.343\,cm^3$ for 31 children (and 59 ear canals) who were between two and ten years old.

25 ear canals have been measured for this thesis. The average mean value of all subjects turned out to be $0.622\,cm^3$. However, for subjects younger than seven years, the mean value is $0.365\,cm^3$, which is very close to the values determined by PFEIL. With advancing age (over seven years) the volume is approx. $0.90\,cm^3$ (see Figure 6.1(a)) which correspond to [SL89]. When taking a closer look at the volume as a function of age, it can be observed that the volume varies from $0.1\,cm^3$ to approx. $0.65\,cm^3$ within the first seven years. This fact underlines that these values are strongly correlated to the growth. Almost no alteration depending on the age can be observed for children who over seven years old.

This is in accordance with the growth of the skull cranial bone. At the age of six, 95% of the growth is completed and at the age of seven, the growing process comes to an end [URL07].

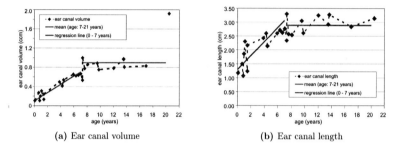

(a) Ear canal volume   (b) Ear canal length

Figure 6.1: Ear canal parameters as a function of age

There is only one outlier in the dataset (age 20.5 years). The ear canal of this subject has an extremely large diameter compared to the other subjects.

During the calculation based on the CAD models some information might get lost, since the CT scans are created with a layer thickness of 0.75 to 1 mm. Therefore the models do not correspond one-to-one.

The correlation coefficient after BRAVIAS-PEARSON is calculated [Bor99] to test the correlation between age and volume. If $r = 1$ there is a perfect correlation between the two values. However, the correlation between two criteria describes a necessary but not sufficient condition for a causal connection. The correlation coefficient for the age group from zero to seven years is $r_{0-7} = 0.949$. This allows the assumption, that the ear canal volume correlates with the age. Figure 6.1(a) shows in addition the regression line (red) for a perfect correlation ($r = 1$). This regression line fits very well to the determined ear canal volumes. Taking a closer look at the subjects who are older than seven years, the correlation coefficient for this group is $r_{7-21} = 0.715$. But without the outlier the $r_{7-20} = 0.298$, which suggests that there is no connection between volume and age.

Ear Canal Length

The determination of the ear canal length turned out to be more difficult. There are different possibilities to determine a curved ear canal. In this thesis the mean value between two splines on the surface is built. The result is similar to the length of the curved central axis of the ear canal. The length of the ear canal is often published with approx. 3.0 cm [BN01]. STINSON ET AL. [SL89] determined values between 3.5 cm and 2.7 cm for subjects with an average age of 55 years for the male and 65 years for the female subjects. STINSON ET AL. determined the length of the curved central axis of the ear canal.

The mean value of the ear canal length is 2.44 cm in this thesis. For the age group of 0–7 years the mean value is 2.03 cm and over seven years it is 2.88 cm. Figure 6.1(b) shows that a similar dependency resembling the one of the volume, can also be established for the relation between the length and the age. The correlation coefficient for subjects under seven years is $r_{0-7} = 0.814$. The regression line for a perfect correlation is very appropriate, although there are some outliers. The correlation coefficient for subjects over seven years is $r_{7-21} = 0.434$ and the ear canal length stabilizes at the age of seven.

Surface Area of the Eardrum

The surface area of the eardrum is published with a total area of 85 mm^2 [BN01]. The mean over all subjects in this thesis is of 52.03 mm^2. On closer inspection this is due to the values from the very young subjects. The eardrum surface areas of children under the age of three are, except one outlier just just between 10 mm^2 and 17 mm^2. The values of the older subjects are more similar to the values published. The mean of the

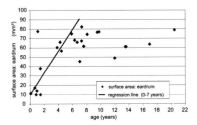

Figure 6.2: Surface area of the eardrum vs. age

subjects older than seven years is 66.1 mm^2. The correlation coefficient for subjects under seven years is $r_{0-7} = 0.763$ (including the outlier). The regression line for a perfect correlation is also plotted in Figure 6.2. In this case greater deviations are obtained. However, a general trend is visible when it comes to the growth dependency of the eardrum surface area. There is no correlation to the age since correlation coefficient for subjects older than seven years is $r_{7-21} = 0.079$.

6.2   Ear Canal Impedances of Children and Adults

All in all 47 ear canal impedances have been studied. 22 children took part in the measurement and 25 CT-scans were available for the simulation.

The age distribution of the measurement subjects are plotted in Figure 6.3(a). For the measurement procedure children needed to be found whose parents gave their consent so that they could take part in the study. The children are students of a special school (and pre-school) for deaf and hearing impaired children. However, there is no reason to consider the ear canals as abnormal. This school was chosen for two reasons. First: the parents need to give their consent to this study. Since individual ear molds are manufactured, where the ear canal is formed with a two-component cold-curing composition, the risks of this procedure were cause of concern for the parents. However, as the parents deal every day with hearing impaired people, they were often willing to give their consent as they were aware of the benefit of this study. The second reason was, that it is a rather awkward feeling when the two-component cold-curing composition is injected into the ear canal during this procedure. Most of the children who are attending this school are wearing hearing aids and are therefore for used to this kind of procedure. These students are undergoing medical treatment regularly and are checked upon every week by a hearing aid consultant. The consultant also knows whether the eardrum and ear canal behaves abnormal or healthy.

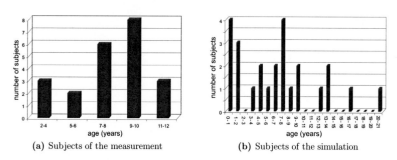

(a) Subjects of the measurement   (b) Subjects of the simulation

Figure 6.3: Age distribution of subjects

It was possible to measure 15 male and seven female subjects in this school. Before the individual ear mold was created, the children's height and age were registered. The measurement was repeated at least four times, with very patient children up to nine times. Each repetition was carried out by taking the ear mold and impedance probe out of the cavum conchae and replacing it again. If a leaky measurement was

found under the repeated measurement, this measurement was rejected and not taken
into account. If the individual ear mold is manufactured with the whole impression
of the cavum conchae up to the pinna fine structure, it is easy to replace the ear mold
at exactly the same position with no degree of freedom.

No problems occurred while using this method on children even when tests were
carried out on three-years-old subjects. If younger children need to be measured, a
smaller probe will have to be used. Since the diameter of the probe would be too
large.

Figure 6.4(a) shows the measured ear canal impedances for all subjects. When
comparing these results with measurement results of CIRIC AND HAMMERSHØI
[CH07], the individual differences among the subjects are in a similar order.

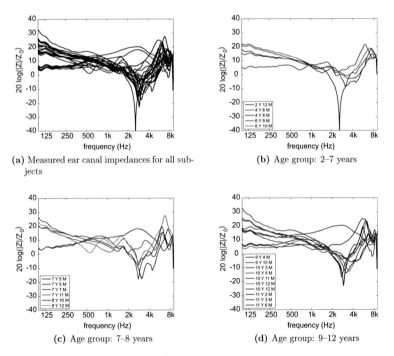

(a) Measured ear canal impedances for all sub-   (b) Age group: 2–7 years
jects

(c) Age group: 7–8 years   (d) Age group: 9–12 years

Figure 6.4: Measured ear canal impedances

The measured impedances of children aged two to seven years are displayed in

Fig. 6.4(b). Compared with the geometrical ear canal parameters, children under the age of six years should show a different ear canal impedance. The first resonance is relatively difficult to identify. The resonance of the youngest child in this survey (two years and 12 months) is at a lower frequency than the ones of the 4.5 years-old subject. The resonance frequencies are slightly higher (around 3.8 kHz) than those obtained for adults.

Fig. 6.4(c) shows the ear canal impedances of seven to eight years-old subjects. The resonance frequencies are around 3 kHz. One child does not fit to the general run of the curves of the others. However, similar outliers are reported by other authors as well ([CH07]).

The age group consisting of children who are over nine years old is shown in Figure 6.4(c). The impedances show an individual deviation, but a dependency on the age can not be observed for subjects of this age.

A correlation between the body height and the ear canal impedance could not be found, since the number of subjects in the age group under the age of six to seven is small.

The test group of the simulation consists of 12 male and 13 female subjects aged between three weeks and 20 years. The focus was eventually put on children under the age of seven for this simulation. The age distribution of the simulation subjects are plotted in Figure 6.3(b).

Figure 6.5 shows the simulated ear canal impedances separately for four age groups. These simulations were carried out without any impedance applied on the eardrum. All boundary conditions are rigid. In comparison to the measurement, again, the simulation shows a considerably higher quality factor.

One can easily see that there is a large variation in the age under 1.5 years (Figure 6.5(a)). The youngest child in this age group is three weeks old. The resonance frequency is at 10.1 kHz and differs clearly from typical adult one's. Even though there are subjects of similar age (e.g. 8, 10, and 11 months) the resonance frequencies differ tremendously with approx. 4 kHz (11 months), 6.8 kHz (8 months) and 9 kHz (10 months). This correlates with the geometrical features of the ear canal. The 8 months-old child has a length of $L = 1.52$ cm and a volume of $V = 0.273$ cm^2. The 10 months-old has a length of $L = 1.08$ cm and a volume of $V = 0.1113$ cm^2 and the 11 months-old subject has a length of $L = 1.87$ cm and a volume of $V = 0.1987$ cm^2.

The same occurs for subjects who are around 1.5 years old. The first resonance of these subjects varies between 2.9 kHz and 9.5 kHz. The child who is 1 year and 1 month old has the lowest first resonance from all 1 year-old children and accordingly this child has the largest ear canal with a length of $L = 2.3$ cm and a volume of $V = 0.31$ cm^2. In contrast to this subject, there is one subject who is 1 year and

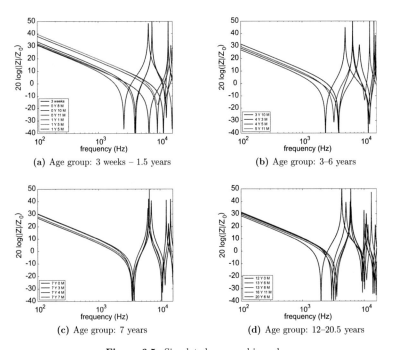

(a) Age group: 3 weeks – 1.5 years (b) Age group: 3–6 years

(c) Age group: 7 years (d) Age group: 12–20.5 years

Figure 6.5: Simulated ear canal impedances

5 months old: The volume is only a third of the 1 year and 1 month old subject ($V = 0.133\,\mathrm{cm}^2$).

Figure 6.5(b) shows the simulated ear canal impedances for four subjects who are between 3 years and 10 months and 5 years and 11 months old. In contrast to the very young subjects, who show a large individual spread, the ear canal impedances of this age group are increasingly similar. The spread is substantially smaller compared to the younger subjects. The first resonance wavers between 2.5 kHz and 5.5 kHz. The geometrical data are also in the same order.

The results for the age group of seven years are more significant (Figure 6.5(c)). The ear canal impedances simulated for the four subjects are very similar. The first resonance is between 3.5 kHz and 4 kHz. The different levels at low frequencies may be caused by the different eardrum surface area. The larger the surface area, the higher is the level at low frequencies. However, the next maxima and minima are not as close together as the first minimum.

The last group consists of all subjects who are over 12 years old. Figure 6.5(d) shows the simulated ear canal impedances. The youngest subject has, however, the longest longest ear canal with a length of $L = 3.25$ cm and a volume of $V = 0.7835$ cm^2. The 20 years-old has the largest volume of all ear canals with $V = 1.925$ cm^2. The length, however, is shorter ($L = 3.15$ cm) than the length of the 12 years-old subject and the eardrum surface area is larger (factor 1.6) which causes the higher first resonance frequency.

When looking at all age groups, one can observe that in general the impedance shows larger values (up to 40 dB at 100 Hz) for very young subjects at very low frequencies.

These simulations were carried out with ideal boundary conditions for the ear canal walls and eardrum.

Influence of the Eardrum Impedance

A lot of different values for the eardrum impedance can be found in relevant literature as it has been pointed out in the previous chapter. The influence of one eardrum impedance after HUDDE [HE98a, HE98b, HE98c] (cf. Section 5.2) is used in the simulations described in this thesis. The ear canal of a 16-years-old male subject is used as an example. The values determined by HUDDE have been fitted with the correct eardrum surface area of 63.6 mm^2. Figure 6.6(a) shows the impedance of this ear canal once calculated with rigid boundary conditions and once with the eardrum impedance applied.

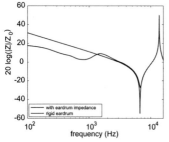

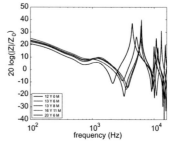

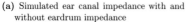

(a) Simulated ear canal impedance with and without eardrum impedance

(b) Age group: 12–20.5 years with eardrum impedance

Figure 6.6: Influence of the eardrum impedance on the ear canal impedance

The difference caused by the eardrum impedance is clearly visible. Only a very slight influence of the eardrum impedance can be detected at frequencies above 4 kHz. The ear canal impedance with eardrum impedance is approx. 10 dB lower than with rigid boundary conditions for frequencies between 100 Hz and 1.0 kHz. At approx. 1.2 kHz the impedance rises to a small maximum and above 1.2 kHz the impedance shows slightly larger values, than without eardrum impedance. Additionally, the quality factor is not as large at the resonance frequencies.

Figure 6.6(b) shows the simulated ear canal impedance subjects who are over 12 years old. One can easily see the influence of the eardrum impedance at frequencies below 4 kHz.

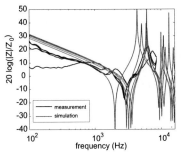 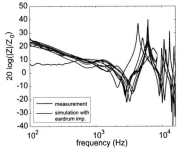

(a) Selection of subjects: measurement results (blue) and simulation results (green) without eardrum impedance

(b) Selection of subjects: measurement results (blue) and simulation results (red) with eardrum impedance

Figure 6.7: Measurement and simulation results; Comparison with and without eardrum impedance applied in the simulation

While comparing the results with the rigid simulations (Fig. 6.5(d)) one can see that the main differences are the alteration in the level at low frequencies (now approx. 22 dB instead of 30 dB), the small maximum at approx. 1.2 kHz, and the lower quality factor.

It was not possible to find a subject for this thesis, who was available for both: measurements and a CT scan. As the risks of a CT-scan are very high, it is not possible to use volunteers. However, if one and the same subject had been available for measurements and simulations, one would certainly have found more details about the damping and other physical issues. This still needs to be done.

However, in Figure 6.7 a selection of measurement results (6, 7, 10 and 11 years-old subjects) are plotted together with simulation results of the group above 12 years.

In (a) the simulation was carried out with rigid boundary conditions and in (b) the eardrum impedance was taken into account.

In Figure 6.7(b) the measurement and simulation results show a very good concordance, although no identical subjects are compared. The typical features are depicted in this case. Using rigid boundary conditions at low frequencies the level is higher than in the measurement and the quality factor is too high. The characteristic maximum which is caused by the eardrum impedance can also be detected in the measurement.

The eardrum impedances, however, are obtained with the help of grown-up subjects. Thus there is a lack of knowledge about the question how eardrum impedances of children differ from adults. It is therefore unclear whether the values determined by HUDDE [HE98a, HE98b, HE98c] are valid for children as well or not.

6.3   Conclusions

The ear canal impedances were measured and simulated for various subjects of different ages. Although the general concordance between these two methods is very satisfying, the simulation probably lacks some physical properties to achieve more realistic results. Properties such as the exact definitions of the ear canal wall properties and the eardrum properties need still to be studied in detail.

However, some general important conclusions can be can be made. The ear canal volume and ear canal length grows from the age of zero to seven years more or less linearly and stagnate above seven years at a certain level. This correlates with the growth of the skull cranial bone. The surface area of the eardrum behaves in a similar manner.

This anatomical features influence the ear canal impedances in a certain way. The ear canal impedances of very young subjects between zero and three years vary significantly, although their anatomical features are quite similar. A small deviation of the eardrum surface area or ear canal length yield in this 'age bracket enormous deviations in the ear canal impedances. In contrast to the three ear canal parameters length, volume, and surface area of the eardrum, the ear canal impedance does not show a dependency to the growth in the first years of life. However, in these first years, the ear canal impedances differ significantly from typical ear canal impedances of adults.

When taking a look at older subjects, however, the growth dependency becomes much clearer. Subjects ranging between three and six years of age, already have similar ear canal impedances within the individual differences (results from simulation and measurement). At the age of seven, the ear canal impedance seems to reach the typical adults characteristics, which is in concordance with the anatomical ear canal parameters.

The ear canal impedances of subjects who are older than seven years old, are already very similar to adults ones.

Part III

Applications and Summary

Chapter 7

Applications

Part I and Part II dealt with the anthropometric data and their influences on binaural cues, HRTFs, and ear canal impedances. In audiology, neuropsychology and psychoacoustics, there are many possible applications for the knowledge about the question of how binaural cues and ear canal impedances grow. For future applications, the results are very important when the focus is put on different generations or age groups, no matter whether the construction of new artificial heads or ears is addressed.

When it comes to *new artificial heads* for children, it depends on the application and the desired accuracy how many heads are needed. Based on the conclusions (Sec. 4.5) of Chapter 4 some suggestions can be stated.

As far as the application of testing of hearing aids with directional signal processing is concerned, it is recommended to create a group of artificial heads (without an ear simulator, since only the directional signal processing is tested) for children under the age of five to six years. Since modern screening technologies allow to provide hearing aids for children at the age of six months, the directional processing should be supported as far as possible.

It is suggested to create at least 2–3 different artificial heads for children for tests or developing hearing aids for children. The first one should focus on infants at an age where hearing aid support begins (which is usually at the age of six months). The second one should focus on children who are two to three years old and the third one should represent an six-years-old child. Above this age, children are easily able to give feedback when it comes to localization tests or hearing aid settings. Thus, it is doubtful whether it would be reasonable to create artificial heads for older children in this field of interest.

There are certainly other applications in the field of hearing impairment of children as well. Virtual scenes with signal and noise setups can be created to test the speech intelligibility in noise for example (cf. Section 7.1). Such scenes can be created by convolving the test signal with the HRTF of the desired angle of incidence and can be presented using headphones or a Cross-Talk-Cancellation.

In this field the subjects are usually older than 3–4 years. Thus, one artificial head of the size of a 5–6 years old would be a good compromise of choice.

Moreover, one important task will be to re-evaluate the standard for artificial heads for adults. It is obvious, when looking at the data suggested in the International Telecommunication Union (ITU) P.58 [ITUP.58] and ITU P.57 [ITUP.57] that this data does not conform with today's values of adults (as already mentioned in Sec. 4.5). Since these datasets are based on anthropometric measurements which were carried out in the 1960s, the anatomy of the population changed during the past 40 years.

The influence of the specific dimensions allows a well-defined procedure of data mining with focused uncertainty budgets. Thus, the most important parameters are known now, databases containing anthropometric data from Europe, America, Australia or Asia (cf. summary Chapter 3, Table 3.2) can be compared to get an overview whether it is possible to create "one" head for all, or whether it is necessary to create various heads due to the differences in the anatomy and the thereby evoked differences in the binaural cues and HRTFs.

The results of the *ear canal impedances* and ear canal parameters can be used as a basis for the development of new suitable couplers for children. Under the age of three years, it is, however, difficult to detect a relation between the age and the ear canal impedances. Thus, creating one coupler for a certain age group is impossible. It is recommended to perform a quick pre-measurement of the ear canal impedance during the fitting process of hearing aids for very young children, to get an indication for the best suitable coupler for testing and fitting purposes and thus calculated the pre-settings for hearing aids without the patient being present.

Between the age of three to seven years the ear canal impedances differ significantly from typical adults ones. This yields a deviation from standard $2\,\mathrm{cm}^3$-coupler-data of about 15–19 dB(SPL) for infants (cf. [Ric80], Section 1.3.2). A correlation between the age and the ear canal impedances can be detected in this period. Thus, is it possible to provide the necessary data for the construction of suitable couplers.

It is possible to construct couplers using various resonators/cavities (similar for example to the ZWISLOCKI COUPLER Section 1.3.2)

At the time being, a few *pilot studies* have been carried out with prototypes of artificial heads for children and these will be described in the following.

7.1   Pilot Study: Hearing Impairment

A study looking at the performance of the *Adaptive Auditory Speech Test* (AAST) [Con06] under spatial locatable noise as a possible diagnostic procedure for *Auditory Processing Disorders* (APD) has been carried out. Auditory processing disorders result from dysfunctions of audition processes and affect the processing of information in the auditory modality.

The AAST was developed to determine the *Speech Recognition Threshold* (SRT) in a fast and reliable way under quiet and noisy conditions for children of 3 to 4 years of age. Usually hearing impaired children take part in this kind of test. Therefore, the test consists of six simple words. The test words are spondee words (metrical foot consisting of two long syllables), which are well known to very young children as well. The words are displayed on a screen. The spondees have a very high redundancy, so that the word can be compared with short sentences. In a closed set procedure the child can click on the word that it understood. If the answer is correct, the next word decreases by 5 dB (under noise condition by 2 dB). If the answer is wrong, the next word will be 10 dB louder (under noise condition 4 dB). The program stops after seven mistakes and the threshold is calculated.

In the study the AAST is used to develop a diagnosis method with a noise made by a talk between children (Two-Talker-Noise) to test the intellectual abilities, separation and localization. The test was carried out under various conditions with the signal coming from the front (0° in the horizontal plane) and noise coming from 0° and 90° in the horizontal plane. The signals were generated using head-related transfer function which are appropriate for this age group.

Therefore, the test words and the two-talker-noise were convolved with the corresponding children HRTFs. The scenarios were presented with the help of headphones.

Significant differences were detected between the situation featuring a signal and a noise coming from the front and a signal coming from the front and noise coming from 90°. If this method can be standardized, this method needs to be analyzed further.

7.2   Pilot Studies: Room Acoustics

In the field of room acoustics in classrooms, appropriate standards and recommendations for the acoustic design are available (cf. e.g. [Bra02], [OT06]). However, numerous schools and teachers complain about poor acoustic conditions. This usually renders the learning process for children more difficult and might lead to dysfunctions of the teacher's voice.

Three pilot studies have dealt with the influence of binaural methods in classroom acoustics. For the first time head-related transfer functions (HRTFs) of children were used for these studies, where several recording methods have been compared in various classroom situations (PRODI ET AL. [PFSF07]), ZHOU [Zho07], and FELS ET AL. [FSV07].

The first study (PRODI ET AL. [PFSF07])) dealt with the loss of Italian language word intelligibility in classrooms caused by a low signal to noise ratio and too high reverberation. Impulse responses and background noises were measured in two primary schools using different mono, binaural and B-format probes. The prototype kindergarten head (cf. Section 4.3.1) was used in this context. It is thus possible to compare the performance of a child head to the conventional adult one. The *Speech Transmission Index* (STI) is calculated under various conditions – with noise from a tapping machine on the upper floor, without noise etc. It turned out that the STI values of the various receivers differ significantly from one another, especially when the S/N is worse.

Similar to this work, the effect of different types of receivers in classrooms on the STI is analyzed. A mono and three binaural receivers (one child head corresponding to the median measures from Section 4.3 and one standardized artificial head, and the adult head corresponding to the measures of GENUIT [Gen84] were compared. The binaural impulse responses of a classroom (see Figure 7.1) were measured and the STI, the *Percentage Articulation Loss of Consonants* (%ALCons), the *Common Intelligibility Scale* (CIS), and the *Clarity Index* (C_{50}) were calculated (see [Zho07] for detailed results). It turned out that the binaural impulse responses are different, especially in octave or third-octave bands where the biggest deviation in the HRTFs of the various heads are.

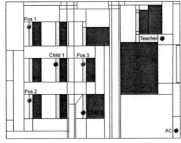

(a) Sketch of the classroom with source and receiver positions

(b) 3D – Model

Figure 7.1: Room acoustic model of the classroom

In contrast to PRODI ET AL. another study (FELS ET AL. [FSV07]) dealt with the HRTFs of children and adults in simulations and evaluated their effects on the speech perception. A typical classroom is modeled corresponding to the dimensions of an original classroom of an elementary school. This classroom model is either empty or occupied. One typical teacher position in front of the blackboard and three receiver positions in the classroom are chosen. Various acoustic scenarios with noise disturbing the teacher's voice are modeled and the Speech Transmission Index is calculated.

To get an impression of the room's influences on speech perception, several experiments were carried out using the room acoustics simulation software RAVEN (Room Acoustics for Virtual Environments) for these simulations. RAVEN uses a hybrid simulation method which combines a deterministic image source method (SCHRÖDER/LENTZ [SL06]) with stochastic ray tracing (SCHRÖDER ET AL. [SDV07]) to compute high quality room impulse responses.

Two models of a typical classroom (volume about 300 m³) were constructed, one representing an empty classroom and one representing an occupied one. The models are built of roughly 180 polygons and 45 planes, respectively. Details of small elements are neglected and represented by equivalent scattering (VORLÄNDER AND MOMMERTZ [VM00]). Moreover, students are also modeled. They are represented by boxes with appropriate surface properties, i.e. absorption and scattering coefficients (ISO 354 [ISO354], ISO/DIS 17497-1 [ISO17497-1]). Figure 7.1(b) shows the model of the occupied classroom. In contrast to the empty room model, it additionally contains black/yellow boxes, which represent the students (note: each color denotes a certain "material"). Room acoustic simulations were performed for various test scenarios, which resulted in a total of 144 computed impulse responses of 2 room models (empty/occupied), 2 HRIR/HRTF databases (child/adult), 12 source/receiver combinations, 3 impulse responses for each combination. The source/receiver positions are displayed in Figure 7.1(a).

A male and a female teacher were chosen to simulate a typical classroom situation for the calculation of the Speech Transmission Index Additionally two different kinds of disturbing noises were generated: speech noise from other students was simulated and an air conditioner was placed in one corner of the room. The source signal of the male and female teacher corresponds to the IEC 60268-16 [IEC60268-16] measurement specification. A level of 65 dB in distance of 1 m from the teacher was chosen for the STI calculation with background noise. This corresponds to normal - slightly raised voice (cf. [ISO9921], [IEC60268-16]). In case of the disturbing noise from two other students, the level was set to be 55 dB in 1 m distance. The signal simulating the children's speech was the female speech. The signal from the air conditioner was measured in the anechoic room with a distance of 1 m.

(a) Child noise 1 to receiver position 3   (b) Noise from AC to receiver position 3

Figure 7.2: Specular reflections from source to receiver (up to third order)

The room impulse responses were simulated for each source and receiver combination. When taking a closer look at the three different receivers, clear differences can be identified, especially, if the impulse responses are third-octave-filtered at 2.7 kHz. This can be put down to the fact that the biggest differences between the two artificial heads exist in this frequency range. As an example, the results at receiver position 3 will be discussed here. Figure 7.2 shows the propagation of strong early specular reflections from child 1 and the air conditioner, respectively, to the receiver at position 3. While the air conditioner is placed in the upper corner of the room, child 1 is located in front of the receiver. This leads to very strong focused specular reflections in case of the AC, and more uniformly distributed specular reflections in case of the child. This yields differences in the respective impulse responses.

Figure 7.3 shows the impulse responses at the exact same position 3 for the adult receiver (left plot) and the child receiver (right plot). The very first reflections are plotted for 2.5 ms. Although the rough structure looks similar, there are tremendous differences in terms of the fine structure of the impulse responses. The lower plot shows a part of the binaural room impulse response, which is third-octave-filtered at 2.7 kHz.

The STI is calculated with the help of the simulated impulse responses for various combinations. In case of the binaural receivers, the highest value was chosen in this evaluation, since there is no recommendation how to calculate the binaural STI.

First of all, the STI is calculated for the empty and occupied room without any influence of noises. The STI was evaluated for receiver position and receiver. In addition, the STI is calculated under different noise conditions. Thus, a female and a male teacher were disturbed by a) child speech from 2 positions and b) by a noisy air conditioner. Figure 7.4 shows the STI values of the three different types of receivers

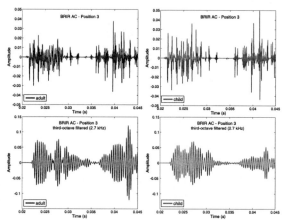

Figure 7.3: Impulse Responses for the path from AC to receiver position 3

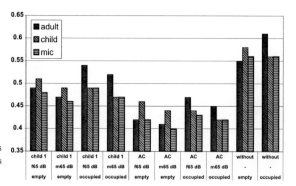

Figure 7.4: STI values for various conditions at receiver position 3

at the receiver position 3 (female and male teacher sound source, empty/occupied classroom condition and configuration with a) noise from child 1 and b) noise from the air conditioner).

Some general results can be stated:

1. Female/male teacher: In all simulated positions and for any kind of receiver the STI-value of the female teacher is approx. 0.2 higher compared to the male teacher.

2. Empty/occupied: whether the room is occupied or empty yields several differences depending on the type of receiver. For the monaural receiver (microphone) the STI is in most cases higher (approx. 0.1) depending on the position. The adult head at receiver position 1 has lower STI values (–0.2) compared to the occupied condition, while at the receiver position 2 and 3 higher values (up to +0.6) can be recognized. In contrast to the child head, this behaves reciprocal to the adult head. At position 1 the STI-values are bigger (+0.1), while at position 2 and 3 the child head produces smaller values (–0.2/ –0.4) in the occupied room condition.

Figure 7.4 also demonstrates how different the various receivers behave in case of noise. The worst Speech Transmission Index occurs in the situation male speaker disturbed by the air conditioner.

Chapter 8

Summary and Outlook

In this thesis, the growth dependency of head-related transfer functions (HRTFs), binaural cues and ear canal impedances are studied. Binaural cues and ear canal impedances of newborn babies up to adults are measured and simulated and the influence of the anthropometric parameters is analyzed.

First, a short introduction is provided and the basics of head-related transfer functions and ear canal impedances are described.

Prior to the studies on head-related transfer functions and ear canal impedances, appropriate methods to determine these attributes for children and infants are developed and tested on children.

Chapter 2 describes a method to determine head-related transfer functions of children. The head-related transfer functions of adults are usually determined by carrying out measurements in an anechoic room. The subject needs to be motionless for a certain period of time. Since such a measurement method cannot be carried out with children easily, an indirect method to obtain HRTFs from children is applied. The HRTFs are simulated numerically using the Boundary Element Method (BEM). The simulation results were compared with ordinary measurement results and showed a very good concordance. CAD models of the head, torso, and pinna are needed for the simulation.

The data acquisition of the anthropometric parameters and the CAD modeling is described in Chapter 3. There are different possibilities to obtain anthropometric parameters, yet, "Photogrammetry" was the method of choice. With a photogrammetric system it is possible to measure any dimension of a head very precisely and at the same time this method provides us with the possibility to use it when it comes to dealing with infants, since this method is contact-free and makes the data collection of movable objects possible. Hence, this system enables us to collect data even for very young children. Using some basic geometrical data, an abstract parametric CAD model of heads can be built. An abstract parametric model was developed, which makes it possible to create abstract models of children. A database was established, containing detailed anthropometric data of 95 subjects ranging between six months and 17.5 years of age.

Since it is impossible to focus on all anthropometric measures and their impact, the attempt was made in this thesis to point out the most important anthropometric parameters. In Chapter 4 the influence of the growth on HRTFs and binaural cues is studied. First, it is studied, whether the head-related transfer functions of a scaled adult head are similar to those of a child head. It turned out that it is not sufficient to approximate children head and torso models with a scaled adult geometry. The scaling of the same head perimeter yield results that are not at all similar to the anatomy of children. This results in different HRTFs and binaural cues, so that it can be concluded that binaural hearing and localization based on these different cues is inappropriate. The differences between a scaled adult and a child are not comparable to the order of individual differences among adults. Using the anthropometric datasets of a six-months-old, a kindergarten child and an adult, one can get an impression how the HRTF, HRIR and binaural cues (ILD and ITD) grow.

The database of anthropometric parameters was reduced to six parameters roughly describing a head: breadth and height of the head, front and back vertex, chin and distance between the ear and the shoulder. Additionally six parameters roughly describing the pinna: breadth and height of outer ear and cavum conchae, depth of the cavum conchae, and rotation of the ear have been chosen. These parameters were studied with regard to their influence on the HRTF and on the binaural cues.

CAD models based on the median data of the database, the 5 %-quantile-values, and the 95 %-quantile-values were built. Moreover, models were created differing from the median head respectively the pinna, only in terms of one parameter. This parameter was changed to the 5 or 95 %-quantile-value. The results showed clearly which parameter brings about which kind of change.

For the head and torso dimensions (Section 4.3), it turned out that the most important parameter with regard to differences in the HRTF and binaural cues is the distance between the ear and the shoulder. It is shown that the breadth of the head and the back vertex have immense influence as well. Only a very slight influence may be attributed to the height of the head, the chin, and the frontal vertex.

As far as the pinna dimensions are concerned (Section 4.4), the parameter varying most of all during growth, i.e. the height of the ear, turned out to have only very little influence on the HRTF and the computed ILD. The greatest influence has to be attributed to the depth and the breadth of the cavum conchae.

The most important anthropometric data is thus available and can be used for future experiments featuring artificial heads and dealing with the special requirements with regard to children. Moreover, the data will be of relevance with regard to new standards for dummy heads.

The ear canal and the corresponding parameters are studied in Chapter 5 and 6.

Chapter 5 describes two methods, which are used to determine the ear canal impedances of children and adults. The first method, which is described, is the measurement of ear canal impedances. A one microphone impedance probe is used to determine the ear canal impedances. The probe is tested and optimized on adults. The coupling of the probe to the ear is accomplished with the help of individual ear molds. This technique yield very reproducible results. Finally, the probe is used for children. The second method is the simulation of ear canal impedances. The simulation is carried out using the Finite-Element-Method (FEM). As for the BEM simulation, a CAD model is needed for the FEM simulation as well. This model is created on the basis of CT-scans of the petrous bone. The transition between air and tissue is clearly visible, hence, in each scan the contour can be traced and finally a 3-dimensional model can be created.

All in all, the ear canal impedances of 47 subjects could be determined. The ear canal impedance of 25 subjects are simulated. The modeling of the ear canal provides a further study of the anatomical parameters of the ear canals.

Chapter 6 described the anthropometric data of the ear canals and its influence on the ear canal impedances. The ear canal parameters lengths, volume, and surface area of the eardrum have been calculated with the help of the CAD models for the 25 subjects aged between newborn and 20.5 years. The three parameters show a strong dependence on the growth during the first six to seven years of life. If the subject is over seven years old, the parameters stagnate at a certain value. However, this strong correlation is not reflected in the same way by simulations of ear canal impedances. First all simulations are carried out with rigid boundary conditions. The ear canal impedances of subjects under three years do not show a dependency on growth. During the first three years the ear canal impedances have a large individual spread. Thus, in this period it is not possible to use the data for uniform input data for couplers for children. For subjects who are over three years old, however, a dependency on growth crystallizes. The values differ from typical adult values as well. With the age of seven years and above the ear canal impedances equal more or less typical adult values. This is in concordance with the anatomical parameters of the ear canal. The simulation is then extended and includes the influence of the eardrum impedance as well. The results differ in a low frequency area from the rigid simulations. However, the values of eardrum impedances are not consistent in relevant literature and for children no data exist.

The measurement results have been obtained for 22 subjects (aged between three years and 11.5 years). Since only three children in this test group are under six years old, the results do not show a strong dependency on growth, since the test subjects are already so old that the ear canal impedances are "fully grown". The measurement and simulation results of comparable age groups show a very good concordance, if the eardrum impedance is taken into account during the simulation.

Outlook   Although all results have a computational character with certain boundary conditions (symmetric geometry, hard surfaces, etc.) and even though a psychoacoustic part is not included, the results can be used as a basis for further studies. One important task will be to re-evaluate standard artificial heads. The influence of the specific dimensions allows a well-defined procedure of data mining with focused uncertainty budgets. Further studies and psychoacoustic experiments will be conducted to create artificial heads for children and adults.

Possible applications are already mentioned in Chapter 7. Some future steps are listed in the following, which will be a dealt with successively before the final applications can be focused on:

- Binaural cues and head-related transfer functions:

    - Collection of more anthropometric data for all age groups (making use of databases, which are already established, too); summarization and evaluation of these datasets to get more precise, reliable and statistically evaluable input data.

    - Listening tests with children of different age groups using suitable artificial heads to test the influence of the perceptional part.

    - Designing of prototypes for artificial heads for children of different age groups in order to test those heads in the field of hearing aid development of children.

    - Simulation of the head-related transfer functions according to the position of hearing aids. Instead of the eardrum the microphone position is used in the simulation.

    - Re-evaluation of standard artificial heads based on the now known important anthropometric data: comparison of the anthropometric data from people from Europe, the Americas, Australia or Asia, to find out whether it is possible to create "one" head for all, or whether it is necessary to create various heads due to the differences in the anatomy and the thereby evoked differences in the binaural cues and HRTFs.

- Ear canal impedances:

  - Simulation and measurement of single subjects, or similar groups of subjects to obtain more precise data of the exact definitions of the ear canal wall properties and the eardrum properties.

  - Determination of other physical factors influencing the results, such as damping or edge effects.

  - Comparison of different measurement methods (for example the latest results by CIRIC AND HAMMERSHØI [CH07]) on the same subject to achieve more exact input data of the real boundary conditions and propagation factors.

  - Construction and testing of prototypes of artificial ears for the fitting process of hearing aids for children according to the ear canal impedances.

Kapitel 9

Kurzfassung

Die Nummerierung der Abschnitte in dieser Kurzfassung entspricht der Nummerierung der Kapitel der Dissertation. Aus Platzgründen ist auf die Darstellung von Bildern verzichtet worden, daher sei zur Erläuterung der Abschnitte auf die Darstellungen in den entsprechenden Kapiteln der Dissertation verwiesen.

Einleitung

Seit den letzten Jahrzehnten werden Kunstköpfe für eine breite Anwendungspalette eingesetzt. Als richtungsabhängiges binaurales (zweiohriges) Mikrophon findet diese Technologie Anwendung in der Audiologie, Raumakustik, Kommunikationstechnik, im akustischen Produktdesign wie auch in der Sound Quality. Kopf-, Torso- und Pinnastrukturen sind dabei die entscheidenden Faktoren der binauralen Aufnahmen.

Zwei Typen von Kunstkopfanwendungen können grundsätzlich unterschieden werden. Bei der Beschallung aus dem Fernfeld werden alle richtungsabhängigen Anteile der Außenohrübertragungsfunktion aufgrund der äußeren Geometrie des Kopfes erfasst. Hierbei wird meist ein Messmikrophon im Gehörgangseingang platziert. Bei kopfnahen Schallquellen ist zusätzlich die Gehörgangsimpedanz, sowie die Cavum Conchae und Pinna von besonderer Bedeutung. In Abhörsituationen werden beide Anteile berücksichtigt: die richtungsabhängigen Anteile durch die Kopf- und Torsogeometrie und unabhängigen Anteile durch den Gehörgang. Eine Auftrennung der beiden Anteile am Gehörgangseingang bietet hierbei eine klar definierte Schnittstelle[1].

---

[1]Bei der Wiedergabe von Kunstkopfaufnahmen über Kopfhörer sind entsprechende Entzerrungen zu beachten.

Internationale Standards schreiben die geometrischen Abmessungen von Kunstköpfen und die akustischen Gegebenheiten von sogenannten Kupplern (Ohrsimulato-
ren) vor. In den letzten Jahren begannen allerdings Diskussionen um die Richtigkeit
dieser Vorschriften. Ein Aspekt ist, dass sich die Anatomie eines „mittleren" Erwachsenen seit der Datenerhebung, auf denen die Standards basieren, verändert hat. Für
Kinder ist die Anwendbarkeit der standardisierten Kunstköpfe überhaupt nicht über-
tragbar, da sich im Laufe des Wachstums Kopf, Torso und Pinna bzw. Gehörgang
unterschiedlich stark verändern. Diese Arbeit untersucht diese Einflüsse und zeigt
wesentliche Konsequenzen für eine zukünftige Entwicklung von Kinderkunstköpfen
und Kupplern für Kinder auf.

9.1   Grundlagen

Das menschliche Gehör ist in der Lage, eine Vielzahl von physikalischen Ereignissen
wahrzunehmen und in ein Raumempfinden und akustische Objekte zu segmentieren.
Aufgrund von Laufzeit- und Pegelunterschieden zwischen den beiden Ohren sowie
durch spektrale Verfärbungen durch Reflexion und Beugung an Kopf, Torso und Pinna
ist der Mensch in der Lage, Geräusche zu lokalisieren.

Die Außenohrübertragungsfunktion (engl. Head-Related Transfer Function
(HRTF)) beschreibt, wie ein Signal auf dem Weg zum Trommelfell aufgrund von Kopf,
Torso und Pinna gefiltert wird. Die Interauralen Zeit- und Pegeldifferenzen (engl. In-
teraural Time Difference (ITD) und Interaural Time Difference (ILD)) können aus der
HRTF abgeleitet werden. Sie bilden den maßgeblichen Satz von Merkmalen, die vom
Gehirn ausgewertet werden und tragen entscheidend zur Lokalisation von Quellen bei.

Kunstköpfe, die die Anatomie des Menschen nachbilden, wurden bereits 1972 für
die Hörgeräteentwicklung entwickelt. Heutzutage finden standardisierte Kunstköpfe
in zahlreichen Arbeitsgebieten ihre Anwendung. Zukünftige Kinderkunstköpfe finden
ihren Einsatz neben der Hörgerätetechnik zum Beispiel auch in der Klassenraumakustik.

Gehörgangsimpedanzen spielen in der Früherkennung von Hörschäden sowie bei
der Anpassung von Hörgeräten eine entscheidende Rolle. Messkuppler bilden den Gehörgang und dessen Eingangsimpedanz nach, um Voreinstellungen am Hörgerät ohne
den Patienten vornehmen zu können. Durch das zu große Volumen in Kupplern, die
heute standardisiert zur Hörgeräteanpassung eingesetzt werden, werden bei Kindern
unangemessene Verstärkungen von bis zu 20 dB(SPL) zu viel eingestellt. Eine Hörgeräteanpassung bei Kindern kann daher heutzutage nur mittels Real-Ear-To-Coupler
Differenzen (RECD) erfolgen. Für die RECD sind zwei Messungen erforderlich. Das

Hörgerät befindet sich in der ersten Messung in der Trageeinstellung und der Schalldruckpegel wird vor dem Trommelfell in dB(SPL) gemessen. Die zweite Messung findet in der Messbox mit Kuppler statt. Die RECD ist die Differenz zwischen den Messboxwerten und den tatsächlichen Pegeln vor dem Trommelfell.

Für Kinder geeignete Kuppler würden eine erhebliche Verbesserung für die Anpassung von Hörgeräten für Kinder erzielen.

Teil I: Außenohr

9.2 Ermittlung von Außenohrübertragungsfunktionen

Grundsätzlich können Außenohrübertragungsfunktionen (HRTFs) auf zwei verschiedene Arten ermittelt werden. In diesem Kapitel werden diese zwei Methoden beschrieben und die Vor- und Nachteile gegeneinander aufgewogen. Die erste und bekannteste Methode ist die Messung von HRTFs auf dem direkten Weg am Individuum. Die zweite Möglichkeit besteht aus der indirekten Ermittlung der HRTF.

Die indirekte Methode setzt jedoch ein Modell von Kopf und Torso voraus. Dieses Modell kann durch das Abformen eines Individuums entstehen, es kann aber auch ein abstraktes Modell anhand von anthropometrischen Daten erzeugt werden. Dies hat ferner den Vorteil, dass auch „mittlere" Kopfabmessungen verarbeitet werden können (Strukturmittelung). Die indirekte Ermittlung der HRTF kann einerseits durch das Messen eines angefertigten Modells oder auch durch die Simulation des Modells erfolgen.

Die Messung von HRTFs erfolgt normalerweise in einer reflexionsfreien Umgebung. Bei der Messung eines Individuums werden am Gehörgangseingang Sondenmikrophone platziert. Ein Lautsprecher sendet aus den gewünschten Schalleinfallswinkeln ein Messignal aus (typischerweise ein deterministisches, breitbandiges Signal, z.B. Sweep). Während dieser Messung darf sich der Proband nicht bewegen. Je nach gewünschter Winkelauflösung dauert eine solche Messung mehrere Stunden. Im Gegensatz zum Erwachsenen ist die Durchführung dieser Methode mit Kindern undenkbar.

Daher wird in dieser Arbeit auf die indirekte Methode zurückgegriffen. Die Ermittlung der HRTFs erfolgt durch Simulation mit Hilfe der Boundary Element Methode

(BEM). Dieses Verfahren ist mittlerweile in der Akustik weit verbreitet. Für eine effiziente Berechnung von kompletten HRTF Datensätzen (in dieser Arbeit 5° über alle Schalleinfallsrichtungen) wird die Simulation reziprok[2] durchgeführt. Durch die reziproke Berechnung muss nur eine Abstrahlanordnung simuliert werden. Die Berechnung der unterschiedlichen Schalleinfallswinkel erfolgt im „Post-Processing" an den gewählten Feldpunkten und zwar quasi simultan.

Diese Vorgehensweise wurde an mehreren Beispielen getestet. Aus CAD-Modellen von Kopf und Torso wurden Modelle gefertigt und die HRTFs dieser Modelle wurden auf die gewöhnliche Weise gemessen. Hierbei wurden in der Messung Mikrophone im Gehörgangseingang platziert. Die Simulationen dieser Modelle wurden reziprok ausgeführt. Der Vergleich der Ergebnisse von Simulation und Messung zeigt eine sehr gute Übereinstimmung.

9.3 Datenerhebung der anthropometrischen Parameter und CAD-Modellierung

Um die Wachstumsabhängigkeit der Außenohrübertragungsfunktion untersuchen zu können, sind die anthropometrischen Parameter von Kopf, Pinna und Torso erforderlich. In diesem Kapitel werden verschiedene Möglichkeiten der Datenerhebung aufgeführt.

In dieser Arbeit wurde die Datenerfassung mit Hilfe der Photogrammetrie durchgeführt. Die Photogrammetrie erlaubt eine berührungslose und kostengünstige Messung beweglicher Objekte. Dadurch ist dieses Verfahren optimal für eine Datenerhebung von Kindern geeignet.

Das Messsystem besteht aus zwei Digitalkameras, die in einem Abstand von ca. 50 cm auf einer Schiene fest montiert sind. Beide Kameras fokussieren so dasselbe Objekt aus zwei leicht unterschiedlichen Perspektiven. Nach Kalibrierung der Kameras und Orientierungsberechnung der Bilder, kann so das photographierte Objekt mit einer Shutterbrille stereoskopisch betrachtet werden und somit dreidimensional ausgewertet werden. Für die Orientierungsberechnung müssen auf jedem der Bilder eindeutige übereinstimmende Punkte gemessen werden. Dies geschieht mit Hilfe von sogenannten Passpunktmarken (ein ca. 3 mm großer schwarzer Kreis), die auf das Gesicht und auf eine Badekappe aufgeklebt sind. Die Badekappe dient zusätzlich dazu, die Haare an die Kopfform anzupressen und so eine genauere Messung zu ermöglichen.

---

[2]Die reziproke Berechnung ist auf Grund der Reziprozität der Green'schen Funktionen möglich: Eine Volumenschnelle (verursacht in der gewöhnlichen Messung durch den Lautsprecher) ruft einen Schalldruck am Gehörgangseingang hervor. Diese Anordnung ist äquivalent zu einer Volumenschnelle am Gehörgangseingang, die einen bestimmten Schalldruck am Ort des Lautsprechers hervorruft.

Es wird jeweils eine Hälfte eines Kopfes vermessen, in dem ein Stereopaar von schräg vorne und ein Stereopaar von hinten angefertigt wird.

Die Probanden sitzen für die Aufnahmen auf einem Stuhl, der auf einem Drehteller steht. Auf diese Weise können die beiden photografierten Perspektiven schnell eingestellt werden. Die Auswertung erfolgt manuell in einer CAD Software mit Photogrammetrie Plug-in. Die Erstellung von CAD-Modellen kann auf zwei verschiedene Weisen erfolgen. Es ist zum einen möglich, ein Modell eines einzelnen Individuums anzufertigen, zum anderen kann auch ein parametrisch abstraktes Modell aufgrund einzelner geometrischer Abmessungen erstellt werden.

Das individuelle Modell ist sehr zeitaufwändig und wegen seiner fein detaillierten Struktur auch sehr rechenintensiv. Um eine Untersuchung der Wachstumsabhängigkeit zu ermöglichen, wird in dieser Arbeit ein parametrisch abstraktes Modell entwickelt, das trotz seiner reduzierten Parameter eine möglichst große Ähnlichkeit zum individuellen Modell aufweist. Ausgehend von einem abstrakten Modell für Erwachsene nach [GENUIT, 1984] wurde ein kindgerechtes Modell entwickelt. Die anthropometrischen Daten, die für dieses Modell erforderlich sind, wurden von 95 Probanden im Alter von sechs Monaten bis zu 18 Jahren ermittelt.

9.4 Analyse der anthropometrischen Daten im Hinblick auf die Außenohrübertragungsfunktion

Während des Wachstums verändern sich die Parameter unterschiedlich. In diesem Kapitel wird der Zusammenhang zwischen den anthropometrischen Daten und der Außenohrübertragungsfunktion (und deren abgeleitete Größen) untersucht.

Zunächst wird allerdings anhand eines Fallbeispiels der Unterschied zwischen einer Kleinkind-HRTF (sechs Monate alt) und eines Erwachsenen gezeigt. Die kompletten Datensätze wurden simuliert und miteinander verglichen. Die HRTFs zeigen eine völlig unterschiedliche Struktur. Des Weiteren wird die Annahme, dass ein skalierter Erwachsenenkopf einem Kinderkopf entspricht, überprüft. Der Erwachsenenkopf wurde auf den gleichen Kopfumfang wie der des sechs Monate alten Kindes herunterskaliert. Diese Skalierung bewirkt in der HRTF eine Frequenzverschiebung zu höheren Frequenzen hin. Dennoch sind die Strukturen der beiden HRTFs sehr unterschiedlich, da ein Kinderkopf eben nicht der Anatomie eines skalierten Erwachsenenkopfes entspricht.

Tabelle 9.1: Eingruppierung der anthropometrischen Parameter des Kopfes anhand des Einflusses auf die HRTF

Abmessung	weniger wichtig	sehr wichtig
1. Abstand: Schulter – Ohr		X
2. Breite		X
3. Scheitelpunkt hinten		X
4. Kinn	X	
5. Scheitelpunkt vorne	X	
6. Kopfhöhe	X	

Die gesammelten anthropometrischen Daten werden statistisch analysiert. Aufgrund der Vielzahl von Probanden im Kindergartenalter wurden die Daten für einen mittleren Kindergartenkopf ermittelt, sowie der Median und Mittelwert und die 5 und 95 %-Quantile der gesamten Datenbasis. Um den Einfluss eines einzelnen Parameters auf die HRTF zu untersuchen, wurde ausgehend von einem Kopf, der aus den Medianwerten erstellt wurde, immer ein Parameter variiert. Dieser Parameter wurde einmal auf den 5 %-Quantilwert verkleinert und einmal auf den 95 %-Quantilwert vergrößert. Durch den Vergleich der HRTFs, ITDs und ILDs der um diesen einen Parameter veränderten Köpfe zum Mediankopf wurde der Einfluss deutlich.

Auf diese Weise wurden die sechs wichtigsten Parameter, die einen Kopf beschreiben (Abstand der Schulter zum Ohr, Kopfbreite, Kopfhöhe, Kinn, Scheitelpunkt vorne (Stirn) und Scheitelpunkt hinten), untersucht. Es zeigte sich, dass die größte Änderung in der HRTF durch die Veränderung des Abstands von Schulter zum Ohr hervorgerufen wird. Ein wesentlicher Einfluss wird auch von der Kopfbreite hervorgerufen. Der Scheitelpunkt hinten zeigt ebenfalls einen Einfluss, während die Parameterveränderungen von Kinn, Kopfhöhe und Scheitelpunkt vorne nur einen sehr geringen Einfluss ausüben.

Entsprechend den Parametervariationen der Kopf- und Torsogeometrie wurde der Einfluss des Wachstums der Pinna auf die HRTF untersucht. Die sieben Parameter Ohrbreite, Ohrhöhe (oben und unten), Cavum Conchae Tiefe, Breite und Höhe sowie die Drehung des Ohres im Verhältnis zum Kopf wurden variiert. Der Parameter, der sich am stärksten während des Wachstums ändert, nämlich die Höhe des Ohres zeigte den geringsten Einfluss auf die HRTF und binauralen Merkmale. Der größte Einfluss wird durch die Breite und Tiefe des Cavum Conchae, sowie durch die Drehung des Ohres hervorgerufen.

Tabelle 9.2: Eingruppierung der anthropometrischen Parameter der Pinna anhand des Einflusses auf die HRTF

Abmessung	weniger wichtig	sehr wichtig
1. Breite (Cavum Conchae)		X
2. Tiefe (Cavum Conchae)		X
3. Drehung		X
4. Höhe (Cavum Conchae)	X	
5. Breite(Pinna)	X	
6. Höhe (Pinna)	X	

Teil II: Gehörgang

9.5 Ermittlung von Gehörgangsimpedanzen

Während in Teil I alle anatomischen Parameter bis zum Gehörgangseingang auf den Einfluss auf die HRTF und binaurale Merkmale durch das Wachstum untersucht worden sind, wird in diesem Teil der Arbeit auf den Gehörgang eingegangen.

Gehörgangsimpedanzen spielen zum Beispiel in der Hörgeräteanpassung sowie bei der Funktionsüberprüfung des Mittelohres eine große Rolle. Zahlreiche Studien bezüglich Gehörgangsimpedanzen, jedoch meist mit Fokus auf die Trommelfellimpedanz, wurden durchgeführt. Nur wenige Untersuchungen beschäftigten sich mit Gehörgangsimpedanzen von Kindern.

Gehörgangsimpedanzen können auf unterschiedliche Art und Weise ermittelt werden. In dieser Arbeit werden zwei unterschiedliche Verfahren angewendet. Ein Messverfahren und ein Simulationsverfahren kommen zum Einsatz. Da es wünschenswert ist, sehr junge Kinder zu vermessen, sollte Messverfahren einerseits eine schnelle Messung ermöglichen und andererseits sollte der Messkopf an das kindliche Ohr angepasst werden können. Es wurde eine Ein-Mikrophon-Messsonde nach dem Prinzip der kalibrierten Quelle gewählt. Die Messsonde besteht aus einem Lautsprecher, der in ein schmales Röhrchen Schall abstrahlt. Am Röhrchen ist seitlich ein Mikrophon angebracht, welches den Schalldruck misst. Mit Hilfe von mindestens zwei Referenzmessungen mit zwei definierten Abschlüssen unterschiedlicher Länge können die Quellparameter der Sonde bestimmt werden. Dann hängt der resultierende Schalldruck nur noch von der angekoppelten Lastimpedanz ab. Somit ist eine Bestimmung der Gehörgangsimpedanz mit nur einer Schalldruckmessung am Gehörgangseingang möglich.

Verschiedene Ankopplungsmöglichkeiten der Sonde an den Gehörgangseingang wurden untersucht. Einfache Gehörschutzstöpsel sowie Tympanometrieoliven zeigten eine hohe Ungenauigkeit und schlechte Reproduzierbarkeit aufgrund der schlechten Abdichtung am Gehörgangseingang. Ferner sollte eine genau definierte Schnittstelle am Übergangsbereich von Cavum Conchae zum Gehörgangseingang vorliegen. Dies kann durch die Verwendung von individuell angefertigten Otoplastiken erzielt werden. Dafür wird eine kalt aushärtende Masse von einem Hörgeräteakustiker in den Gehörgang eingebracht und bis zum Cavum Conchae aufgefüllt. Gleichzeitig wird ein Stift in der Größe der Sonde mit abgeformt, so dass nach dem Aushärten die Sonde exakt in dieses Loch eingesetzt werden kann. Die Ankopplungsebene kann aufgrund des Abdrucks genau bestimmt werden. Um Übergang von Cavum Conchae zum Gehörgang wird die Otoplastik abgeschnitten und somit wird die Messebene vordefiniert. Die Wiederholbarkeit von Messungen mit derselben Otoplastik sowie mit mehreren angefertigten Otoplastiken wurde am Erwachsenen untersucht.

Das zweite verwendete Verfahren zur Bestimmung der Gehörgangsimpedanz ist die Simulation mit Hilfe der Finite Elemente Methode (FEM). Auch hier bildet die Grundlage der Simulation ein CAD-Modell. Um reale Gehörgänge vom Kleinkind bis hin zum Erwachsenen untersuchen zu können, wurden die Gehörgangsmodelle mit Hilfe von CT-Bildern des Felsenbeins erstellt. Der Übergang von Luft zu Gewebe ist in diesen Bildern deutlich zu erkennen, auch die Lage des Trommelfells ist gut sichtbar. In jeden Schnittbild werden die Konturen des Gehörgangs in einem CAD Programm nachgezeichnet. Somit kann aus diesen Konturen ein dreidimensionales Modell erstellt werden. Auch hier wird die Eingangsebene am Übergang zwischen Cavum Conchae und Gehörgang festgelegt. Die Randbedingungen des Modells können frei gewählt werden – schallhart oder mit einem Impedanzbelag. Die Finite Elemente Methode berechnet den Schalldruck am Gehörgangseingang bei eingeprägter Schnelle. Die Gehörgangsimpedanz kann so bestimmt werden.

Diese beiden Verfahren wurden anhand eines Probevolumens miteinander verglichen. Das Probevolumen in der Größe eines kleineren Erwachsenen Gehörgangs wurde einmal aus Aluminium gefertigt und einmal als CAD-Modell erstellt. Der Vergleich der Ergebnisse von Simulation und Messung liefert sehr zufriedenstellende Ergebnisse.

9.6   Analyse der Gehörgangsparameter im Hinblick auf die Gehörgangsimpedanzen

In diesem Kapitel werden zunächst die Gehörgangsparameter ausgewertet und analysiert. Ferner wird die Wachstumsabhängigkeit der gemessenen und simulierten Gehörgangsimpedanzen untersucht.

Für die Simulation standen 25 CT-Daten von Probanden im Alter von drei Wochen und 20,5 Jahren zur Verfügung. Aufgrund der CAD-Modelle ist es möglich Gehörgangsparameter abzuleiten. Die Parameter Volumen, Länge und die Fläche des Trommelfells wurden berechnet und deren Wachstumsabhängigkeit untersucht. Die Parameter Gehörgangsvolumen und -länge zeigten identische Merkmale. Eine starke Korrelation zum Alter zeigte sich in den ersten sieben Lebensjahren. Dabei korreliert das Gehörgangsvolumen noch stärker mit dem Alter als die Gehörgangslänge. Ab sieben Jahren stagniert das Wachstum von Volumen und Gehörgangslänge. Ähnlich verhält sich die Fläche des Trommelfells. Auch hier ist eine Korrelation in den ersten sieben Lebensjahren zu erkennen, allerdings mit wesentlich stärkerer Streuung.

Insgesamt konnten durch Messungen und Simulationen die Gehörgangsimpedanzen von 47 Probanden ermittelt werden. 22 Probanden nahmen an Messungen teil. Die Altersspanne der gemessenen Gehörgangsimpedanzen liegt zwischen drei und zwölf Jahren. Das gewählte Messverfahren bereitete auch bei den jüngsten Probanden keine Probleme.

Durch Simulation wurden 25 weitere Gehörgangsimpedanzen bestimmt. Eine große Streuung zeigte sich vor allem bei den Gehörgangsimpedanzen der sehr jungen Probanden unterhalb von zwei Jahren, allerdings ohne einen direkten Zusammenhang zum Alter. Ab sieben Jahren liegen die Gehörgangsimpedanzen im Bereich eines Erwachsenen.

Die Simulationen wurden zusätzlich zur schallharten Randbedingung auch unter Berücksichtigung von Trommelfellimpedanzen durchgeführt. Diese Ergebnisse zeigten eine große Ähnlichkeit zu den Messergebnissen.

Teil III: Anwendungen und Zusammenfassung

9.7 Anwendungen

Teil I und II beschäftigten sich mit den anthropometrischen Daten und deren Auswirkungen auf binaurale Merkmale, HRTFs sowie Gehörgangsimpedanzen. Es bestehen zahlreiche Anwendungsmöglichkeiten durch das Wissen wie HRTFs und Gehörgangsimpedanzen wachsen. Die Ergebnisse sind für zukünftige Anwendungen mit Hauptaugenmerk auf unterschiedliche Altersgruppen sehr wichtig.

Bei der Neuentwicklung von Kunstköpfen für Kinder hängt es von der Art der Anwendung ab, wie viele verschiedene Köpfe nötig sind. Für die Entwicklung von Hörgeräten sollten beispielsweise mehrere Kinderkunstköpfe für Kinder unterhalb von 5-6 Jahren verwendet werden, da sich in dieser Zeit die wichtigen Parameter sehr stark verändern. Heutzutage werden moderne Hörgeräte mit Richtwirkung schon mit sechs Monaten angepasst, um somit eine optimale Versorgung (z.B. im Hinblick auf die Sprachentwicklung) zu gewährleisten.

Es wurden im Rahmen dieser Arbeit mehrere Pilotstudien mit Prototypen von Kinderkunstköpfen durchgeführt. Um die Sprachverständlichkeit von hörgeschädigten Kindern einer virtuellen Szene mit Störgeräusch zu testen, wurden die Außenohrübertragungsfunktionen von Kindern im entsprechenden Alter mit Nutz- und Störsignalen aus den gewünschten Richtungen gefaltet. So können unterschiedliche Szenarien dargeboten und der Einfluss auf räumlich lokalisierbare Quellen untersucht werden.

Mehrere Untersuchungen wurden mit Prototypen von Kinderkunstköpfen im Bereich der Klassenraumakustik durchgeführt. Der Kinderkunstkopf wurde im Vergleich zu standardisierten Kunstköpfen und Messmikrophonen zur Messung von raumakustischen Parametern und zur Bestimmung der Sprachverständlichkeit eingesetzt. Hierbei wurden Messungen in Klassenräumen vorgenommen, sowie Simulationen der Raumimpulsantworten durchgeführt. Die Unterschiede der verschiedenen Messmethoden sind deutlich erkennbar, allerdings kann kein allgemeines Ergebnis abgeleitet werden. Besonders unter dem Einfluss von Störgeräuschen (wie bei Lüfter- oder Klimaanlagen zum Beispiel) ergeben sich große Differenzen durch die verschiedenen Kopfabmessungen.

Die Ergebnisse der Gehörgangsimpedanzen können als Datenbasis für die Entwicklung von neuen kindgerechten Kupplern dienen. Auch hier muss der Fokus auf die Altersgruppe unterhalb von 6-7 Jahren gelegt werden. Hier weichen die Impedanzen deutlich von Erwachsenen ab.

9.8 Zusammenfassung

In dieser Arbeit werden die Wachstumsabhängigkeiten von Außenohrübertragungsfunktionen und daraus abgeleiteten Größen (Interaurale Zeit- und Pegelunterschiede) sowie von Gehörgangsimpedanzen untersucht. Für die Untersuchung an Kindern entwickelte Verfahren machen es möglich, die anthropometrischen Parameter vom Kleinkind bis hin zum Erwachsenen zu erfassen. Mittels geeigneter Simulations- und Messverfahren werden diese Daten hinsichtlich ihres Einflusses auf binaurale Merkmale und auf Gehörgangsimpedanzen statistisch ausgewertet.

Es zeigt sich, dass die Außenohrübertragungsfunktion eines Kindes wesentliche Unterschiede zur der eines Erwachsenen aufweist. Ferner kann eine Außenohrübertragungsfunktion eines Kindes nicht durch das Skalieren von Erwachsenen-Abmessungen angenähert werden. Durch die unterschiedliche Detail-Anatomie zwischen Kind und Erwachsenen entstehen große Unterschiede in den binauralen Merkmalen. Allerdings wirken sich die einzelnen anthropometrischen Parameter unterschiedlich stark auf die binauralen Merkmale aus. In dieser Arbeit werden die wichtigsten anthropometrischen Parameter hinsichtlich des Einflusses auf binaurale Größen ermittelt. Auch die Gehörgangsimpedanzen ändern sich stark im Laufe des Wachstums vom Kleinkind bis hin zu Erwachsenen. Die Arbeit zeigt erstmals den Verlauf der altersabhängigen Entwicklung der für die Impedanz maßgeblichen Daten.

Mit Hilfe der gewonnenen Ergebnisse stehen nun neue Möglichkeiten für die Entwicklung von Kinderkunstköpfen und kindgerechten Hörgerätekupplern zur Verfügung. Dadurch können spezielle Kinderanwendungen, wie zum Beispiel die Hörgeräteentwicklung und -anpassung oder auch die Messmethodik in der Klassenraumakustik wesentlich optimiert werden. Ferner dienen die gewonnenen Erkenntnisse der Re-Evaluierung von standardisierten Kunstköpfen.

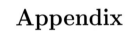

Appendix

Glossary

Curriculum Vitae

Personal Data

Janina Fels

05.04.1977   born in Mönchengladbach, Germany

Education

08/1983–06/1987   Primary School, "Katholische Grundschule Niederkrüchten"

08/1987–06/1996   Secondary School, "Gymnasium St. Wolfhelm", Schwalmtal

Course of Studies

10/1996–03/2002   Study of Electrical Engineering at the RWTH Aachen University

09/2001–03/2002   Diploma Thesis at the Institute of Technical Acoustics,
                   RWTH Aachen University

since 04/2002   Ph.D. at the Institute of Technical Acoustics,
                   RWTH Aachen University

Employments

10/2000–02/2001   Internship at Westdeutscher Rundfunk (WDR), Köln

04/2001–07/2001   Internship at RPG Diffusor Systems, Inc., Upper Marlboro, USA

since 04/2002   Research Assistant at the Institute of Technical Acoustics,
                   RWTH Aachen University

Danksagung

An der Entstehung dieser Arbeit war eine Vielzahl von Personen direkt und indirekt beteiligt, denen ich im folgenden meinen Dank aussprechen möchte. Für die Betreuung dieser Arbeit danke ich Herrn Prof. Dr. rer. nat. Michael Vorländer, ohne dessen Übersicht, Motivation und hilfreiche Diskussionen diese Arbeit nicht möglich gewesen wäre.

Für die freundliche Übernahme des Korreferats und die ungewöhnlich schnelle Durchsicht der Arbeit danke ich Herrn Prof. Dr.-Ing. Peter Vary.

Allen Mitarbeitern des Instituts für Technische Akustik danke ich für die Unterstützung und das wunderbare Arbeitsklima. Besonders hervorheben möchte ich Dipl.-Ing. Andreas Franck und Dr.-Ing. Michael Makarski, die mich auf den Wegen in die Welt der Numerik hilfreich unterstützten. Den Werkstätten des Instituts möchte ich meinen Dank aussprechen für die Herstellung von etlichen Prototypen von Kunstköpfen sowie Messaufbauten. Meinen Diplomanden Dipl.-Ing. Pia Buthmann, Dipl.-Ing. Jan Paprotny, Dipl.-Ing. Lars Feickert und Dipl.-Ing. Mei Zhou sowie meinen studentischen Hilfskräften Martin Peter und David Hense sei gedankt für die wertvollen Arbeiten.

Herrn Dipl.-Ing. Christoph Effkemann danke ich für die tatkräftige Hilfe mit der Photogrammetrie-Software.

Prof. Dr.-Ing. Herbert Hudde und Dipl.-Ing. Sebastian Schmidt von der Ruhr-Universität Bochum danke ich für die hilfreichen Diskussionen über die Messtechnik der Impedanzmesssonde.

Herrn Matthias Kaulard danke ich für die Unterstützung bei der Anfertigung der Otoplastiken für die Impedanzmessungen.

Ein herzlicher Dank geht an Dr. med. Justus Illgner und Dr.-Ing. Wolfgang Döring von der Klinik für Hals-, Nasen-, Ohrenheilkunde und Plastische Kopf- und Halschirurgie sowie Dr. med. Gabriele Krombach und Dr. med. Georg Mühlenbruch von der Klinik für Radiologische Diagnostik des Universitätsklinikums Aachen für die bereitgestellten CT-Bilder.

Der Kooperationsbereitschaft der Leiter, Eltern und Kinder der Familienbildungsstätte Aachen, Liebfrauenkindertagesstätte und -hort Mönchengladbach, Evangelischen Integrativen Kindertagesstätte Kaiserpark in Schwalmtal, Katholischen Tageseinrichtung für Kinder Himmelsschaukel in Mönchengladbach, David-Hirsch Schule für Hörgeschädigte Aachen, Montessori Grundschule Reumontstr. Aachen und Viktoriaschule Aachen sei an dieser Stelle nochmals herzlich gedankt.

Vielen Dank an Stephanie Heikamp, die auf wunderbare und schnelle Art mein Englisch korrigierte.

Zu guter Letzt danke ich meiner Familie für die stetige Unterstützung nicht nur im Rahmen dieser Arbeit.

Bibliography

[AAD01a] V. R. Algazi, C. Avendano and R. O. Duda, *Estimation of a spherical-head model from anthropometry*, Journal of the Audio Engineering Society **49** (2001), no. 6, 472–478.

[AAD01b] V. R. Algazi, C. Avendano and R. O. Duda, *Elevation localization and head-related transfer function analysis at low frequencies*, The Journal of the Acoustical Society of America **109** (2001), no. 3, 1110–1122.

[ADD⁺02] V. R. Algazi, R. O. Duda, R. Duraiswami, N. A. Gumerov and Z. Tang, *Approximating the head-related transfer function using simple geometric models of the head and torso*, The Journal of the Acoustical Society of America **112** (2002), no. 5, 2053–2064.

[ADMT01] V. R. Algazi, R. O. Duda, R. P. Morrison and D. M. Thompson, *Structural Composition and Decomposition of HRTFs*, 2001 IEEE ASSP Workshop on Applications of Signal Processing to Audio and Acoustics, WASSAP '01 (Mohonk Mountain House, New Paltz, NY, USA), October 2001, pp. 103–106.

[ADTA01] V. R. Algazi, R. O. Duda, D. M. Thompson and C. Avendano, *The CIPIC HRTF database*, 2001 IEEE ASSP Workshop on Applications of Signal Processing to Audio and Acoustics, WASSAP '01 (Mohonk Mountain House, New Paltz, NY, USA), October 2001, pp. 99–102.

[AL68] M. Alexander and L. L. Laubach, *Anthropometry of the human ear. (A photogrammetric study of USAF flight personnel)*, Tech. Report AMRL-TR-67-203, Air Force Systems Command, Wright-Patterson AB, OH, 1968.

[ANSIS3.25] ANSI S3.25, *Occluded Ear Simulator*, American National Standard ANSI S3.25-1989 (R2003), American National Standards Institute, 1989.

[Bar03]   G. Bartsch, *Effiziente Methoden für die niederfrequente Schallfeldsim-ulation*, Ph. D. thesis, Institut für Technische Akustik, RWTH Aachen University, 2003.

[Bat67]   D. W. Batteau, *The Role of the Pinna in Human Localization*, Proceedings of the Royal Society of London. Series B, Biological Sciences, vol. 168, 1967, pp. 158–180.

[Bat68]   D. W. Batteau, *Listening with the naked ear*, Chapter 7 in: The Neuropsychology of Spatially Oriented Behavior (Homewood, Illinois) (S. Freedman, ed.), Dorsey Press, 1968, pp. 109–133.

[BD98]   C. P. Brown and R. O. Duda, *A structural model for binaural sound synthesis*, IEEE Transactions on Speech and Audio Processing **6** (1998), no. 5, 476–488.

[BE91]   W. Benning and C. Effkemann, *PHIDIAS - ein photogrammetrisch interaktives digitales Auswertesystem für den Nahbereich.*, Zeitschr. f. Photogrammetrie und Fernerkundung **59** (1991), 87–93.

[Bék32]   G. Békésy, *Über den Einfluß der durch den Kopf und den Gehörgang bewirkten Schallfeldverzerrungen auf die Hörschwelle*, Annalen der Physik **406** (1932), no. 1, 51–56.

[Bla74]   J. Blauert, *Räumliches Hören*, ISBN 3-7776-0250-7, S. Hirzel Verlag Stuttgart, 1974.

[Bla97]   J. Blauert, *Spatial Hearing: The Psychophysics of Human Sound Localization*, revised ed., ISBN 0-262-02413-6, Cambridge, Massachusetts, London: The MIT Press, 1997.

[BM71]   A. J. Burton and G. F. Miller, *The appliction of integral methods to the solution of some exterior boundaryvalue problems*, Proceedings of the Royal Society of London **A** (1971), no. 323, 201–210.

[BN01]   H. Brunner and I. Nöldeke, *Das Ohr*, 2 ed., Thieme, Stuttgart; New York, 2001.

[Bor99]   B. Bortz, *Jürgen: Statistik für Sozialwissenschaftler*, Springer Verlag, Berlin, 1999.

[BP76]   J. Blauert and H. J. Platte, *Impulsmessung der menschlichen Trommelfellimpedanz*, Zeitschrift für Hörgeräte Akustik **3** (1976), 34–44.

[BR88]   I. Brand and L. Reinken, *Diagramme: Wachstums- und Gewichtskurven in Perzentilen*, Klinische Pädiatrie **200** (1988), 451–456.

[Bra02]   J. Bradley, *Acoustical Design of Rooms for Speech*, Construction Tech-
               nology Update 51, Institute for Research in Construction, 2002.

[BS75]   M. D. Burkhard and R. M. Sachs, *Anthropometric manikin for acoustic
               research*, The Journal of the Acoustical Society of America **58** (1975),
               no. 1, 214–222.

[BS97]   W. Benning and R. Schwermann, *PHIDIAS.MS - eine digitale Pho-
               togrammetrieapplikation unter MicroStation für Nahbereichsanwen-
               dungen (PHIDIAS.MS - a digital photogrammetry tool for near
               field applications using Microstation)*, Allgemeine Vermessungs-
               Nachrichten (AVN) **104** (1997), 16–25.

[Bur78]   M. Burkhard (ed.), *Manikin Measurements*, Elk Grove Village, Illi-
               nois, USA, (A Knowles Company), Industrial Research Products Inc.,
               February 1978, now GRAS.

[BV04]   G. K. Behler and M. Vorländer, *Reciprocal measurements on con-
               denser microphones for quality control and absolute calibration*, ACTA
               ACUSTICA united with ACUSTICA **90** (2004), no. 1, 152–160.

[CH06]   D. G. Ciric and D. Hammershøi, *Coupling of earphones to human ears
               and to standard coupler*, The Journal of the Acoustical Society of
               America **120** (2006), no. 4, 2096–2107.

[CH07]   D. Ciric and D. Hammershøi, *Acoustic impedances of ear canals mea-
               sured by impedance tube*, The 19th International Congress on Acous-
               tics, ICA 2007 (Madrid, Spain), September 2007.

[Con06]   F. Coninx, *Entwicklung und Erprobung des Adaptiven Auditiven
               Sprachtests (AAST), (Development and Evaluation of the adaptive
               auditory speech test (AAST))*, Deutsche Gesellschaft für Audiologie:
               Neunte Jahrestagung, 2006.

[CT56]   E. Churchill and B. Truett, *Metrical Relations Among Dimensions Of
               The Head And Face*, Technical Report WADC TR 56-621, Wright Air
               Development Center, 1956.

[DIN33402-1]   DIN 33402-1, *Körpermaße des Menschen; Begriffe, Meßverfahren,
               (Body dimensions of people; terms and definitions, measuring proce-
               dures)*, Norm DIN 33402-1, Deutsches Institut für Normung, January
               1978.

[DIN33402-2]   DIN 33402-2, *Ergonomie – Körpermaße des Menschen – Teil 2:
               Werte, (Ergonomics – Human body dimensions – Part 2: Values)*,
               Norm DIN 33402-2, Deutsches Institut für Normung, December 2005.

[Dre67]   H. Dreyfuss, *The Measure of Man*, Whitney Library of Design, New York, 1967.

[Fah95]   F. Fahy, *The Vibro-Acoustic Reciprocity Principle and Applications to Noise Control*, Acustica **81** (1995), 544–558.

[Fas04]   H. Fastl, *Towards a New Dummy Head?*, The 33rd Intern. Congress on Noise Control Engineering, INTER-NOISE 2004 (Prague, Czech Republic), August 2004.

[FBV04]   J. Fels, P. Buthmann and M. Vorländer, *Head-related transfer functions of children*, ACTA ACUSTICA united with ACUSTICA **90** (2004), no. 5, 918–927.

[FDF⁺07]   T. Fedtke, P. Daniel, H. Fastl, T. Fedtke, K. Genuit, H. P. Grabsch, T. Niederdränk, A. Schmitz, M. Vorländer and M. Zollner, *Kunstkopftechnik - Eine Bestandsaufnahme; Eine Mitteilung aus dem Normenausschusses "Psychoakustische Messtechnik" (NA 001-01-02-08 AK, vormals NALS A 2 AK 8)*, ACUSTICA/acta acustica/Nuntius Acusticus CD-ROM **93** (2007), no. 1, 1–58.

[Fed07]   T. Fedtke, *Kunstkopftechnik - Eine Bestandsaufnahme*, report, Eine Mitteilung aus dem Normenausschusses "Psychoakustische Messtechnik" (NA 001-01-02-08 AK, vormals NALS A 2 AK 8), 2007.

[Fei07]   L. Feickert, *Messung von Gehörgangsimpedanzen, (Measurement of ear canal impedances)*, Master's thesis, Institute of Technical Acoustics, RWTH Aachen University, Germany, 2007.

[Fel05]   J. Fels, *The Influence of Different Head Geometries on Spatial Hearing*, The 2005 International Congress on Noise Control Engineering, Inter-Noise 2005 (Rio de Janeiro, Brazil), 2005.

[FFSR67]   H. Fischler, E. H. Frei, D. Spira and M. Rubinstein, *Dynamic Response of Middle-Ear Structures*, The Journal of the Acoustical Society of America **41** (1967), no. 5, 1220–1231.

[Fra03]   A. Franck, *Finite Element formulations for structural acoustics*, The 32nd International Congress on Noise Control Engineering – Inter-Noise 2003 (Jeju, Korea), August 25-28 2003.

[FSV07]   J. Fels, D. Schröder and M. Vorländer, *Room acoustics simulations using head-related transfer functions of children and adults*, International Symposium on Room Acoustics, ISRA 2007 (Seville, Spain), 2007.

[Gen84]   K. Genuit, *Ein Modell zur Beschreibung von Aussenohrübertragungs-seigenschaften (A model for describing transfer-functions of the outer ear)*, Ph. D. thesis, Institut für Technische Akustik, RWTH Aachen University, 1984.

[GF07]   K. Genuit and A. Fiebig, *Do we need new artificial heads?*, The 19th International Congress on Acoustics, ICA 2007 (Madrid, Spain), September 2007.

[GS02]   R. Gan and Q. Sun, *Finite element modeling of human ear with external ear canal and middle ear cavity*, Engineering in Medicine and Biology, 2002. 24th Annual Conference and the Annual Fall Meeting of the Biomedical Engineering Society; EMBS/BMES Conference, 2002. Proceedings of the Second Joint, vol. 1, 2002.

[HE98a]   H. Hudde and A. Engel, *Measuring and modeling basic properties of the human middle ear and ear canal. Part I: Model structure and measuring techniques*, Acustica united with Acta Acustica **84** (1998), 720–738.

[HE98b]   H. Hudde and A. Engel, *Measuring and modeling basic properties of the human middle ear and ear canal. Part II: Ear canal, middle ear cavities, eardrum, and ossicles*, Acustica united with Acta Acustica **84** (1998), 894–913.

[HE98c]   H. Hudde and A. Engel, *Measuring and modeling basic properties of the human middle ear and ear canal. Part III: Eardrum impedances, transfer functions and model calculations*, Acustica united with Acta Acustica **84** (1998), 1091–1109.

[Hel83]   R. Helle, *Abschätzung der wirksamen Verstärkung des Hörgerätes am Ohr des Kindes*, Pädaudiologie aktuell, 27 Vorträge des pädaudiologischen Symposiums am 14. und 15. Oktober 1983 in Mainz (P. D. med. Peter Biesalski, ed.), Universitätsdruckerei und Verlag Dr. Hanns Krach Mainz, 1983.

[HM96]   D. Hammershøi and H. Møller, *Sound transmission to and within the human ear canal*, The Journal of the Acoustical Society of America **100** (1996), no. 1, 408–427.

[Hud83]   H. Hudde, *Measurement of the eardrum impedance of human ears*, The Journal of the Acoustical Society of America **73** (1983), no. 1, 242–247.

[IEC60711]   IEC 60711, *Occluded-ear simulator for the measurement of earphones coupled to the ear by ear inserts*, International Standard IEC 60711, International Electrotechnical Commission, January 1981.

[IEC60118-0]   IEC 60118-0, *Hearing aids; Part 0: Measurement of electroacoustical characteristics (IEC 60118-0:1983 + IEC 118-0 AMD 1:1994)*, International Standard IEC 60118-0, International Electrotechnical Commission, 1983.

[IEC60959]   IEC 60959, *Provisional head and torso simulator for acoustic measurements on air conduction hearing aids (will become IEC 60318-7)*, International Standard IEC 60959 TR:1990, International Electrotechnical Commission, 1990.

[IEC60268-16] IEC 60268-16, *Sound system equipment – Part 16: Objective rating of speech intelligibility by speech transmission index*, International Standard IEC 60268-16, International Electrotechnical Commission, 2003.

[IEC60118-7]   IEC 60118-7, *Electroacoustics – Hearing aids – Part 7: Measurement of performance characteristics of hearing aids for production, supply and delivery quality assurance purposes*, International Standard IEC 60118-7, International Electrotechnical Commission, 2005.

[IEC60118-8]   IEC 60118-8, *Electroacoustics – Hearing aids – Part 8: Methods of measurement of performance characteristics of hearing aids under simulated in situ working conditions*, International Standard IEC 60118-8, International Electrotechnical Commission, 2005.

[IEC60318-5]   IEC 60318-5, *Electroacoustics – Simulators of human head and hear – Part 5: 2cm3 coupler for the measurement of hearing aids andearphones coupled to the ear by means of ear inserts (This first edition of IEC 60318-5 cancels and replaces IEC 60126:1973)*, International Standard IEC 60318-5, International Electrotechnical Commission, 2006.

[ISO354]   ISO 354, *Acoustics – Measurement of sound absorption in a reverberation room*, Standard ISO 354, International Organization for Standardization, 2003.

[ISO9921]   ISO 9921, *Ergonomics – Assessment of speech communication*, Standard ISO 9921, International Organization for Standardization, 2003.

[ISO17497-1]   ISO 17497-1, *Acoustics – Sound-scattering properties of surfaces – Part 1: Measurement of the random-incidence scattering coefficient in a reverberation room*, Standard ISO 17497-1, International Organization for Standardization, 2004.

[ITUP.57]   ITU-T P.57, *Artificial Ears*, Recommendation ITU-T P.57, International Telecommunication Union, March 1993.

[ITUP.58]   ITU-T P.58, *Head and Torso Simulator for Telephonometry*, Recommendation ITU-T P.58, International Telecommunication Union, March 1993.

[Joh87]   C. Johnson, *Numerical solution of partial differential equations by the finite element method*, Cambridge University Press New York, 1987.

[Kat00]   B. F. G. Katz, *Acoustic absorption measurement of human hair and skin within the audible frequency range*, The Journal of the Acoustical Society of America **108** (2000), no. 5, 2238–2242.

[Kat01]   B. F. G. Katz, *Boundary element method calculation of individual head-related transfer function. I. Rigid model calculation*, The Journal of the Acoustical Society of America **110** (2001), no. 5, 2440–2448.

[KBAB93]   D. H. Keefe, J. C. Bulen, K. H. Arehart and E. M. Burns, *Ear-canal impedance and reflection coefficient in human infants and adults*, The Journal of the Acoustical Society of America **94** (1993), no. 5, 2617–2638.

[Kel85a]   F. Keller, *F.K.K.K. Freiburger Konischer Kinder-Kuppler; Erstmals möglich: Kleinkinderbezogene Messung und Auswahl von Hörgeräten mit etymotischen Korrekturen*, Audiotechnik **35** (1985).

[Kel85b]   F. Keller, *Vorschlag eines konischen Kupplers; Ausführung für Kleinkinder*, Audio-Technik **35** (1985), 32–36, Vortrag, gehalten auf dem Jubiläumskongreß der AKUSTIKA am 5. Juni 1983 in Zürich.

[KNPC98]   Y. Kahana, P. A. Nelson, M. Petyt and S. Choi, *Boundary Element Simulation of HRTFs and Sound Fields Produced by Virtual Acoustic Imaging Systems*, Proceedings of the 105th Audio Engineering Society Convention, August 1998.

[KR87]   B. Kruger and R. Ruben, *The acoustic properties of the infant ear. A preliminary report.*, Acta Otolaryngol. **103** (1987), no. 5-6, 578–585.

[Kuh77]   G. F. Kuhn, *Model for the interaural time differences in the azimuthal plane*, The Journal of the Acoustical Society of America **62** (1977), no. 1, 157–167.

[KWK02]   T. Koike, H. Wada and T. Kobayashi, *Modeling of the human middle ear using the finite-element method*, The Journal of the Acoustical Society of America **111** (2002), no. 3, 1306–1317.

[LA97]   R. Y. Litovsky and D. H. Ashmead, *Development of Binaural and Spatial Hearing in Infants and Children*, Chapter VII.27 in: Binaural and Spatial Hearing in Real and Virtual Environments (Hillsdale, New Jersey, USA) (R. H. Gilkey and T. R. Anderson, eds.), ISBN 0-8058-1654-2, Lawrence Erlbaum Associates, 1997, pp. 571–591.

[LH85]   U. Letens and H. Hudde, *Akustische Impedanzmessung im Gehörgang unter Ausschluss des Einflusses inhomogener Querschnittsverläufe*, Fortschritte der Akustik - DAGA 1985, 1985, pp. 755–758.

[LH95]   A. Lodwig and H. Hudde, *Akustische Impedanzmessungen am Ohr durch Otoplastiken*, Fortschritte der Akustik - DAGA 1995 (Saarbrücken), 1995, pp. 223–226.

[LNCJE93]   V. Larson, J. Nelson, W. Cooper Jr and D. Egolf, *Measurements of acoustic impedance at the input to the occluded ear canal*, Journal of Rehabilitation Research and Development **30** (1993), 129–129.

[Lya59]   L. M. Lyamshev, *A question in connection with the principle of reciprocity in acoustics*, Soviet Physics Doklady **4** (1959), 406.

[Mec02]   F. P. Mechel, *Formulas of Acoustics*, Springer Verlag, 2002.

[MG91]   J. C. Middlebrooks and D. M. Green, *Sound localization by human listeners*, Annual Review of Psychology **42** (1991), 135–159.

[MHJS99]   H. Møller, D. Hammershøi, C. B. Johnson and M. F. Sørensen, *Evaluation of Artificial Heads in Listening Tests*, Journal of the Audio Engineering Society **47** (1999), no. 3, 83–100.

[Mid99a]   J. C. Middlebrooks, *Individual differences in external-ear transfer functions reduced by scaling in frequency*, The Journal of the Acoustical Society of America **106** (1999), no. 3, 1480–1492.

[Mid99b]   J. C. Middlebrooks, *Virtual localization improved by scaling nonindividualized external-ear transfer functions in frequency*, The Journal of the Acoustical Society of America **106** (1999), no. 3, 1493–1510.

[Mil58]   A. W. Mills, *On the Minimum Audible Angle*, The Journal of the Acoustical Society of America **30** (1958), no. 4, 237–246.

[MJ56]   J. Y. Morton and R. A. Jones, *The Acoustical Impedance Presented by Some Human Ears to Hearing-Aids Earphones for the Insert Type*, Acustica **6** (1956), 339–349.

[MM77]   S. Mehrgardt and V. Mellert, *Transformation characteristics of the external human ear*, The Journal of the Acoustical Society of America **61** (1977), no. 6, 1567–1576.

[MPO⁺00]   P. Minnaar, J. Plogsties, S. K. Olesen, F. Christensen and H. Møller, *The interaural time difference in binaural synthesis*, Proceedings of the 108th Audio Engineering Society Convention (Paris, France), preprint 5133, February 2000, pp. 1–20.

[MSHJ95]   H. Møller, M. F. Sørensen, D. Hammershøi and C. B. Jensen, *Head-related transfer functions of human subjects*, Journal of the Audio Engineering Society **43** (1995), no. 5, 300–321.

[MSS94]   K. Moodie, R. Seewald and S. Sinclair, *Procedure for Predicting Real-Ear Hearing Aid Performance in Young Children*, American Journal of Audiology **3** (1994), no. 1, 23–31.

[Onc61]   Y. Onchi, *Mechanism of the Middle Ear*, The Journal of the Acoustical Society of America **33** (1961), no. 6, 794–805.

[OT06]   M. Oberdörster and G. Tiesler, *Acoustic ergonomics of school - a premise for "modern teaching"?*, Euronoise 2006 (Tampere, Finland), 2006.

[Pfe79]   H. Pfeil, *Die Bedeutung der Form des kindlichen Gehörganges für pädaudiolgische Diagnostik und Therapie*, Sprache - Stimme - Gehör **3** (1979), 134–140.

[PFSF07]   N. Prodi, A. Farnetani, Y. Smyrnova and J. Fels, *Investigating Classroom Acoustics by means of Advanced Reproduction Techniques*, Proceedings of the 122th Audio Engineering Society Convention (Vienna, Austria), May 2007.

[PLMI89]   A. Prader, R. Largo, L. Molinari and C. Issler, *Physical growth of Swiss children from birth to 20 years of age; First Zurich Longitudinal Study of Growth and Development*, Helvetica Paediatrica Acta **Suppl. 52** (1989), 1–125.

[Rab81]   W. M. Rabinowitz, *Measurement of the acoustic input immittance of the human ear*, The Journal of the Acoustical Society of America **70** (1981), no. 4, 1025–1035.

[Ray07] L. Rayleigh, *On our perception of sound direction*, Philos. Mag **13** (1907), 214–232.

[RDBP46] F. E. Randall, A. Damon, R. S. Benton and D. I. Patt, *Human Body Size in Military Aircraft and Personal Equipment*, Technical Report 5501, Army Air Force, Air Material Command, Wright Field, Dayton, OH, June 1946.

[Ric80] U. Richter, *Kinderohr-Simulator*, Tech. Report 3.5.10, PTB Jahresbericht 1980, 1980, p. 189.

[Rie03] K. A. J. Riederer, *Effects of eye-glasses, hair, headgear, and clothing on measured head-related transfer functions Part Ib*, The Journal of the Acoustical Society of America **114** (2003), no. 4, 2388–2388.

[SD05] M. R. Stinson and G. A. Daigle, *Comparison of an analytic horn equation approach and a boundary element method for the calculation of sound fields in the human ear canal*, The Journal of the Acoustical Society of America **118** (2005), no. 4, 2405–2411.

[SDV07] D. Schröder, P. Dross and M. Vorländer, *A Fast Reverberation Estimator for Virtual Environments*, AES 30th International Conference, Saariselkä, Finland, 2007.

[Sha66] E. A. G. Shaw, *Earcanal Pressure Generated by a Free Sound Field*, The Journal of the Acoustical Society of America **39** (1966), no. 3, 465–470.

[Sha76] E. A. G. Shaw, *Diffuse field sensitivity of external ear based on reciprocity principle*, The Journal of the Acoustical Society of America **60** (1976), no. S1, S102–S102.

[Sha97] E. A. G. Shaw, *ch. I.2. Acoustical Features of the Human External Ear*, in: Binaural and Spatial Hearing in Real and Virtual Environments (Hillsdale, New Jersey, USA) (R. H. Gilkey and T. R. Anderson, eds.), ISBN 0-8058-1654-2, Lawrence Erlbaum Associates, 1997, pp. 25–48.

[SL89] M. R. Stinson and B. W. Lawton, *Specification of the geometry of the human ear canal for the prediction of sound-pressure level distribution*, The Journal of the Acoustical Society of America **85** (1989), no. 6, 2492–2503.

[SL06] D. Schröder and T. Lentz, *Real-Time Processing of Image Sources Using Binary Space Partitioning*, Journal of the Audio Engineering Society **54** (2006), no. 7/8, 604–619.

[SSO⁺77]   R. G. Snyder, L. W. Schneider, C. L. Owings, H. M. Reynolds,
           D. H. Golomb and M. A. Schork, *Anthropometry of infants, children
           and youths to age 18 for product safety design.*, Final report UM-
           HSRI-77-17, Biomedical Department, Highway Safety Research Insti-
           tute, Ann Arbor, Mich., 1977.

[SSOS75]   R. G. Snyder, M. L. Spencer, C. L. Owings and L. W. Schneider, *Phys-
           ical characteristics of children as related to death and injury for con-
           sumer product safety design. (also published by SAE, titled "Anthro-
           pometry of U.S. infants and children.")*, Final report UM-HSRI-BI-75-
           5, Biomedical Department, Highway Safety Research Institute, Ann
           Arbor, Mich., 1975.

[SSOVE72] R. G. Snyder, M. Spencer, C. Owings and P. Van Eck, *Source data
           of infant and child measurements.*, Interim data UM-HSRI-BI-72-3,
           Biomedical Department, Highway Safety Research Institute, Ann Ar-
           bor, Mich., 1972.

[ST68]   E. A. G. Shaw and R. Teranishi, *Sound Pressure Generated in an
           External-Ear Replica and Real Human Ears by a Nearby Point Source*,
           The Journal of the Acoustical Society of America **44** (1968), no. 1,
           240–249.

[Sti90]   M. R. Stinson, *Revision of estimates of acoustic energy reflectance at
           the human eardrum*, The Journal of the Acoustical Society of America
           88 (1990), no. 4, 1773–1778.

[Str73]   J. W. Strutt, *Some general theorems relating to vibrations*, Proc. Lon-
           don Math. Soc., no. 4, 1873, pp. 357–368.

[SWM02]   K. Stephan and K. Welzl-Müller, *Diagnostik von Hörstörungen mit
           objektiven Verfahren: Impedanzaudiometrie*, Deutsche Gesellschaft für
           Audiologie: Fünfte Jahrestagung, 2002.

[SWW06]   J. Saffran, J. Werker and L. Werner, *The Infant's Auditory World:
           Hearing, Speech, and the Beginnings of Language*, Chapter 2 in:
           6th Edition of the Handbook of Child Development (R. Siegler and
           D. Kuhn, eds.), New York: Wiley, 2006, pp. 58–108.

[URL07]   URL, *http://www.uke.uni-hamburg.de/
           kliniken/neurochirurgie/index_15719.php*, visited September, 2007.

[VDH⁺06]   L. Vallejo, V. Delgado, A. Hidalgo, E. Gil-Carcedo, L. Gil-Carcedo and
           F. Montoya, *Modelling of the geometry of the external auditory canal*

by the finite elements method, Acta Otorrinolaringol Esp **57** (2006), 82–89.

[VM00]   M. Vorländer and E. Mommertz, *Definition and measurement of random-incidence scattering coefficients*, Applied Acoustics **60** (2000), no. 2, 187–199.

[WK92]   F. L. Wightman and D. J. Kistler, *The dominant role of low-frequency interaural time differences in sound localization*, The Journal of the Acoustical Society of America **91** (1992), no. 3, 1648–1661.

[ZAC60]   R. Zeigen, M. Alexander and E. Churchill, *A head circumference sizing system for helmet design, including three-dimensional presentation of anthropometric data*, Technical Report WADC-TR-60-631, Wright Air Development Center, Wright-Patterson Air Force Base, 1960.

[Zho07]   M. Zhou, *Untersuchung der Akustik von Klassenräumen mit binauralen Methoden, (Investigation of classroom acoustics with binaural methods)*, Master's thesis, Institute of Technical Acoustics, RWTH Aachen University, Germany, 2007.

[Zie77]   O. C. Zienkiewicz, *The Finite Element Method*, 3rd ed., McGraw-Hill, London, 1977.

[Zwi70]   J. J. Zwislocki, *An acoustic coupler for earphone calibration.*, Special Report LSC-S-7, Laboratory of Sensory Communication, Syracuse University, Syracuse, New York, 1970.

[Zwi71]   J. J. Zwislocki, *An earlike coupler for earphone calibration.*, Special Report LSC-S-9, Laboratory of Sensory Communication, Syracuse University, Syracuse, New York, 1971.